Student Solutions Manual
for Rice and Strange's

College Algebra

with Applications

FOURTH EDITION

Carroll M. Schleppi
University of Dayton

Brooks/Cole Publishing Company
Pacific Grove, California

Brooks/Cole Publishing Company
A Division of Wadsworth, Inc.

Printed in the United States of America

10   9   8   7   6   5   4   3   2   1

ISBN 0-534-10208-5

Sponsoring Editors: Faith Stoddard and Sue Ewing
Production: Dorothy Bell
Cover Design: Flora Pomeroy

An Introduction for Students

Every odd-numbered problem with an answer that requires some work to reach a solution has been solved in this manual. The method used in each case matches the method used in that section of the text. Some simple steps have been excluded.

Use this manual to check your solutions or to assist you when you are having difficulty solving a problem. Be aware that often several correct methods for doing a problem may be used. The method you select may be different from the one included in the manual, but also correct. It is unwise to use the manual as a substitute for solving the problems in the text.

If you have difficulty following any of the solutions provided, please feel free to contact me c/o Brooks/Cole Publishing Company, 511 Forest Lodge Road, Pacific Grove, California 93950.

Carroll M. Schleppi

University of Dayton

Contents

Chapter 1 Basic Algebra

Section 1.1 Real Numbers

21. $-2 - (7-10)(5) = -2 - (-3)(5) = -2 + 15 = 13$

23. $-\big[2 + 3(6-8)\big]\big[4 - (5+2)\big] = -\big[2 + 3(-2)\big]\big[4-7\big] = -\big[2 + 3(-2)\big]\big[-3\big] = -\big[-4\big]\big[-3\big]$
    $= -12$

25. $\big[-(-2) - (4-6)\big]\big[3 + (7-3)\big] = \big[2-(-2)\big]\big[3 + 4\big] = [4][7] = 28$

27. $-2\big[7.1 + (1.2-2.8) - 6(2.5-3.2)\big] = -2\big[7.1 - 1.6 - 6(-.7)\big] = -2\big[5.5+4.2\big] = -2\big[9.7\big]$
    $= -19.4$

51. $|6-9| = |-3| = 3$ 53. $5 + |-9| = 5 + 9 = 14$

55. $3 + |-3| = 3 + 3 = 6$ 57. $1 - |3| = 1 - 3 = -2$

59. $d = |-3 - 1| = 4$ 61. $d = |-5 - (-1)| = |-5 + 1| = 4$

63. $d = |-0.1 - 7| = 7.1$ 65. $d = |-6 - 0| = 6$

67. $d = |-0.125 - 0.176| = 0.301$

69. $|a+b| \neq |a| + |b|$

    Try $a = 5,\ b = -3$

    $|5-3| = 2 \neq |5| + |-3| = 8$

    $|a+b| = |a| + |b|,$ when $a \geq 0$ and $b \geq 0$
     or $a < 0$ and $b < 0$

Section 1.2 Integral Exponents

21. $\dfrac{7x^2 y^5}{27xy} = \dfrac{7}{27}xy^4$ 23. $\dfrac{(x^2)^6}{(y^5)^6} = \dfrac{x^{12}}{y^{30}}$

29. $(x^{-2}y^5)^4 = \left(\dfrac{y^5}{x^2}\right)^4 = \dfrac{y^{20}}{x^8}$ 31. $\left(\dfrac{y}{2x}\right)^3 = \dfrac{y^3}{8x^3}$

33. $\left(\dfrac{a^{-3}b^7}{a^2 b^4}\right)^{-2} = \left(\dfrac{b^3}{a^5}\right)^{-2} = \left(\dfrac{a^5}{b^3}\right)^2 = \dfrac{a^{10}}{b^6}$

35. $\left(\dfrac{x^3 y^{-6}}{x^7 y^{-4}}\right)^{-2} = \left(\dfrac{1}{x^4 y^2}\right)^{-2} = \left(x^4 y^2\right)^2 = x^8 y^4$

39. $x^0 + (2y)^0 = 1 + 1 = 2$

57. $(x+y)^{-1} \overset{?}{=} x^{-1} + y^{-1}$

    Consider $x = 2,\ y = 3$

$$(2+3)^{-1} = \tfrac{1}{5}$$

$$2^{-1} + 3^{-1} = \tfrac{1}{2} + \tfrac{1}{3} = \tfrac{5}{6}$$

    Therefore

$$(x+y)^{-1} \neq x^{-1} + y^{-1}$$

59. $(x^2)^3 \overset{?}{=} x^{2^3}$

    Consider $x = 2$

$$(2^2)^3 = 4^3 = 64$$

$$2^{2^3} = 2^8 = 256$$

    So $(x^2)^3 \neq x^{2^3}$

Section 1.3   Radicals and Fractional Exponents

1. $\sqrt{49} = \sqrt{7^2} = 7$

3. $\sqrt[3]{125} = \sqrt[3]{5^3} = 5$

5. $\sqrt[5]{-32} = \sqrt[5]{(-2)^5} = -2$

7. $\sqrt[7]{5^{14}} = \sqrt[7]{(5^2)^7} = 5^2 = 25$

9. $5\sqrt{243} = 5\sqrt{9^2 \cdot 3} = 5 \cdot 9\sqrt{3} = 45\sqrt{3}$

11. $\sqrt{a^4 b^2} = \sqrt{(a^2)^2 b^2} = a^2 b$

13. $\sqrt{63a^5 b^8} = \sqrt{7}\sqrt{3^2}\sqrt{(a^2)^2\, a}\sqrt{(b^4)^2} = 3a^2 b \cdot \sqrt[4]{7a}$

15. $\sqrt[4]{2x^3 z^{12}} = \sqrt[4]{2}\,\sqrt[4]{x^3}\,\sqrt[4]{(z^3)^4} = z^3\,\sqrt[4]{2x^3}$

17. $2m\sqrt{m^5 n^3} = 2m\sqrt{(m^2)^2 m}\sqrt{n^2 n} = 2m \cdot m^2 n\sqrt{mn} = 2m^3 n\sqrt{mn}$

19. $\sqrt{\dfrac{x^7}{y^4}} = \dfrac{\sqrt{(x^3)^2 x}}{\sqrt{(y^2)^2}} = \dfrac{x^3\sqrt{x}}{y^2}$

21. $\dfrac{2}{\sqrt{5}} \cdot \dfrac{\sqrt{5}}{\sqrt{5}} = \dfrac{2\sqrt{5}}{5}$

23. $\dfrac{1}{\sqrt{a}} \cdot \dfrac{\sqrt{a}}{\sqrt{a}} = \dfrac{\sqrt{a}}{a}$

25. $\sqrt{5}\sqrt{20} = \sqrt{100} = 10$

27. $\sqrt[3]{2a}\,\sqrt[3]{4a^2} = \sqrt[3]{8a^3} = \sqrt[3]{2^3}\,\sqrt[3]{a^3} = 2a$

35. $\sqrt{x\sqrt{x\,\sqrt[3]{x}}} = \left(x\left(x(x^{1/3})\right)^{1/2}\right)^{1/2} = \left(x\left(x^{4/3}\right)^{1/2}\right)^{1/2} = \left(x \cdot x^{2/3}\right)^{1/2} = \left(x^{5/3}\right)^{1/2} = x^{5/6}$

39. $\left(a^{3/2}\right)^{1/4} = a^{3/8} = \sqrt[8]{a^3}$

45. $81^{3/4} = \left(81^{1/4}\right)^3 = 3^3 = 27$

47. $(32)^{-3/5} = \left(32^{1/5}\right)^{-3} = 2^{-3} = \dfrac{1}{8}$

49. $x^{1/2} x^{2/3} = x^{1/2\,+\,2/3} = x^{7/6}$

51. $y^{4/3} y^{-1/2} = y^{(4/3)(-1/2)} = y^{5/6}$

53. $(125a^3 b^9)^{1/3} = 125^{1/3} ab^3 = 5ab^3$

55. $\left(9x^{2/3}\right)^{1/2} = 9^{1/2} x^{1/3} = 3x^{1/3}$

57. $\sqrt[4]{y}\,\sqrt[3]{y} = y^{1/4} y^{1/3} = y^{1/4\,+\,1/3} = y^{7/12} = \sqrt[12]{y^7}$

59. $\sqrt[3]{5}\,\sqrt[4]{4} = 5^{1/3} 4^{1/4} = 5^{4/12} 4^{3/12} = (5^4 4^3)^{1/12} = \sqrt[12]{5^4 4^3}$

61. $\sqrt{x}\,\sqrt[3]{(x+y)} = x^{1/2}(x+y)^{1/3} = x^{3/16}(x+y)^{2/6} = \left(x^3 (x+y)^2\right)^{1/6} = \sqrt[6]{x^3 (x+y)^2}$

63. $\sqrt{5\sqrt{x^4}} = \left(x^{4/5}\right)^{1/2} = x^{2/5} = \sqrt[5]{x^2}$ 65. $27^{1/6} = \left(27^{1/3}\right)^{1/2} = \sqrt{3}$

67. $\sqrt[4]{(a+1)^2} = (a+1)^{2/4} = \sqrt{a+1}$ 69. $\sqrt[6]{(y+2)^3} = (y+2)^{3/6} = \sqrt{y+2}$

79. $v = kT^{1.2} = (2.3 \times 10^4)(13^{1.2}) = 4.994 \times 10^5$

81. $\left(a^{2/3}+b^{2/3}\right)^{3/2} = \left(1.3^{2/3}+2.5^{2/3}\right)^{3/2} = 5.28$

$\left(\sqrt{a}+\sqrt{b}\right)^2 = \left(\sqrt{1.3}+\sqrt{2.5}\right)^2 = 7.41$

Section 1.4 Multiplication of Algebraic Expressions

1. $x^2(x^3+2) = x^2x^3 + x^2(2) = x^5 + 2x^2$

3. $x^3(x^2-5x) = x^3x^2 + x^3(-5x) = x^5 - 5x^4$

5. $(x+2y)(x-2y) = x(x-2y) + 2y(x-2y) = x^2 - 2xy + 2xy - 4y^2 = x^2 - 4y^2$

7. $(x^2-1)(x^2+1) = x^2(x^2+1) - (x^2+1) = x^4 + x^2 - x^2 - 1 = x^4 - 1$

9. $\sqrt{x}(\sqrt{x}+\sqrt{y}) = \sqrt{x}\sqrt{x} + \sqrt{x}\sqrt{y} = x + \sqrt{xy}$

11. $(xy+1)(xy-1) = xy(xy-1) + (xy-1) = x^2y^2 - xy + xy - 1 = x^2y^2 - 1$

13. $(x+y^2)^3 = \left[(x+y^2)(x+y^2)\right](x+y^2)$

$= \left[x(x+y^2) + y^2(x+y^2)\right](x+y^2)$

$= \left[x^2+xy^2+xy^2+y^4\right](x+y^2)$

$= \left[x^2+2xy^2+y^4\right]x + \left[x^2+2xy^2+y^4\right]y^2$

$= x^3 + 2x^2y^2 + xy^4 + x^2y^2 + 2xy^4 + y^6$

$= x^3 + 3x^2y^2 + 3xy^4 + y^6$

15. $(x+y)(x-y)(x^2-y)$

$= \left[x(x-y) + y(x-y)\right](x^2-y)$

$= \left[x^2-xy+xy-y^2\right](x^2-y)$

$= (x^2-y^2)x^2 + (x^2-y^2)(-y)$

$= x^4 - x^2y^2 - x^2y + y^3$

17. $(x+y+z)(x+y+z) = x(x+y+z) + y(x+y+z) + z(x+y+z)$

$= x^2 + xy + xz + xy + y^2 + yz + xz + yz + z^2$

$= x^2 + 2xy + 2xz + 2yz + y^2 + z^2$

19. $(3x+4y-z)(2x-3y) = 3x(2x-3y) + 4y(2x-3y) - z(2x-3y)$

$\qquad = 6x^2 - 9xy + 8xy - 12y^2 - 2xz + 3yz$

$\qquad = 6x^2 - 12y^2 - xy - 2xz + 3yz$

27. $(2x+5)^2 = (2x)^2 + 2(2x)5 + 5^2 = 4x^2 + 20x + 25$

29. $(\sqrt{a}-\sqrt{b})^2 = (\sqrt{a})^2 - 2\sqrt{a}\sqrt{b}+(\sqrt{b})^2 = a - 2\sqrt{ab} + b$

31. $(4x+5)(4x-5) = (4x)^2 - 5^2 = 16x^2 - 25$

35. $(5a+b)^3 = (5a)^3 + 3(5a)^2 b + 3(5a)b^2 + b^3 = 125a^3 + 75a^2 b + 15ab^2 + b^3$

37. $(2x-5)^3 = (2x)^3 - 3(2x)^2 5 + 3(2x)5^2 - 5^3 = 8x^3 - 60x^2 + 150x - 125$

43. $(\sqrt{C}+\sqrt{D})(C-\sqrt{CD}+D) = (\sqrt{C})^3 + (\sqrt{D})^3 = C\sqrt{C} + D\sqrt{D}$

45. $(x+5)(x+9) = x^2 + 9x + 5x + 45 = x^2 + 14x + 45$

47. $(2x-8)(3x+5) = 6x^2 + 10x - 24x - 40 = 6x^2 - 14x - 40$

49. $(2y+5)(y+2) = 2y^2 + 4y + 5y + 10 = 2y^2 + 9y + 10$

51. $(2x-3y)(x+4y) = 2x^2 + 8xy - 3xy - 12y^2 = 2x^2 + 5xy - 12y^2$

53. $(12x-5)(3x-8) = 36x^2 - 96x - 15x + 40 = 36x^2 - 111x + 40$

55. $\dfrac{(2+h)^3 - 8}{h} = \dfrac{8 + 3(4)h + 3(2)h^2 + h^3 - 8}{h} = \dfrac{\cancel{h}(12+6h+h^2)}{\cancel{h}} = 12 + 6h + h^2$

57. $(5t-3)(2t-1) = 10t^2 - 5t - 6t + 3 = 10t^2 - 11t + 3$

59. $4(2T+3)(3T+1) = 4(6T^2+2T+9T+3) = 4(6T^2+11T+3) = 24T^2 + 44T + 12$

Section 1.5   Factoring

5. $z^2 - 24z + 144 = z^2 - 24z + 12^2 = (z-12)^2$

7. $9x^2 + 6x + 1 = (3x)^2 + 6x + 1 = (3x+1)^2$

9. $\frac{1}{4}x^2 + x + 1 = (\frac{x}{2})^2 + x + 1 = (\frac{x}{2}+1)^2$

11. $x^2 - 169 = x^2 - 13^2 = (x-13)(x+13)$

13. $4m^2 - 64 = 4(m^2-16) = 4(m^2-4^2) = 4(m-4)(m+4)$

15. $x^2 - 7 = x^2 - (\sqrt{7})^2 = (x-\sqrt{7})(x+\sqrt{7})$

17. $\frac{1}{4}x^2 - \frac{1}{9} = (\frac{1}{2}x)^2 - (\frac{1}{3})^2 = (\frac{1}{2}x-\frac{1}{3})(\frac{1}{2}x+\frac{1}{3})$

19. $9x^2 + 12x + 4 = (3x)^2 + 12x + 2^2 = (3x+2)^2$

21. $\frac{1}{9}x^2 + \frac{2}{9}x + \frac{1}{9} = \frac{1}{9}(x^2+2x+1) = \frac{1}{9}(x+1)^2$

39. $8x^2 - 20xy + 8y^2 = 4(2x^2-5xy+2y^2) = 4(2x-y)(x-2y)$

41. $x^3 + 64 = x^3 + 4^3 = (x+4)(x^2-4x+16)$

43. $8x^3 + 1 = (2x)^3 + 1 = (2x+1)(4x^2-2x+1)$

45. $a^3 + \frac{1}{8} = a^3 + \left(\frac{1}{2}\right)^3 = \left(a+\frac{1}{2}\right)\left(a^2-\frac{a}{2}+\frac{1}{4}\right)$

47. $y^3 - 9y^2 + 27y - 27 = y^3 - 9y^2 + 27y - 3^3 = (y-3)^3$

49. $8x^3 + 12x^2 + 6x + 1 = (2x)^3 + 12x^2 + 6x + 1 = (2x+1)^3$

51. $a^2 + a^3b + ab^3 + b^2 = a^2(1+ab) + b^2(ab+1) = (1+ab)(a^2+b^2)$

53. $z^2 - w^2 - 3z + 3w = (z-w)(z+w) - 3(z-w) = (z-w)(z+w-3)$

55. $a^2 - c^2 + 2cd - d^2 = a^2 - (c^2-2cd+d^2) = a^2 - (c-d)^2$

$= \left[a-(c-d)\right]\left[a+(c-d)\right] = (a-c+d)(a+c-d)$

57. $x^2+4x+1 = x^2 + 4x + 4 - 4 + 1$

$= x^2 + 4x + 4 - 3$

$= (x+2)^2 - 3$

$= (x+2-\sqrt{3})(x+2+\sqrt{3})$

59. $a^2-8a-5 = a^2 - 8a + 16 - 16 - 5$

$= (a-4)^2 - 21 = (a-4-\sqrt{21})(a-4+\sqrt{21})$

61. $x^2+3x+\frac{3}{4} = x^2 + 3x + \frac{9}{4} - \frac{9}{4} + \frac{3}{4}$

$= \left(x+\frac{3}{2}\right)^2 - \frac{3}{2} = \left(x+\frac{3}{2}-\sqrt{\frac{3}{2}}\right)\left(x+\frac{3}{2}+\sqrt{\frac{3}{2}}\right)$

63. $4y^2-8y-8 = 4(y^2-2y+1) - 4 - 8$

$= 4(y-1)^2 - 12 = \left[2(y-1)\right]^2 - (\sqrt{12})^2$

$= \left[2(y-1)-\sqrt{12}\right]\left[2(y-1)+\sqrt{12}\right]$

$= \left[2(y-1)-2\sqrt{3}\right]\left[2(y-1)+2\sqrt{3}\right]$

$= 4(y-1-\sqrt{3})(y-1+\sqrt{3})$

65. $2b^2+6b-3 = 2\left(b^2+3b+\frac{9}{4}\right) - \frac{9}{2} - 3$

$= 2\left(b+\frac{3}{2}\right)^2 - \frac{15}{2} = 2\left(b+\frac{3}{2}\right)^2 - \frac{15(2)}{4}$

$= 2\left(b+\frac{3}{2}-\frac{\sqrt{15}}{2}\right)\left(b+\frac{3}{2}+\frac{\sqrt{15}}{2}\right)$

67. $x^2+Lx-2L^2 = \left(x^2+Lx+\frac{L^2}{4}\right) - \frac{L^2}{4} - 2L^2$

$$= \left(x+\frac{L}{2}\right)^2 - \frac{9L^2}{4} = \left(x+\frac{L}{2}-\frac{3L}{2}\right)\left(x+\frac{L}{2}+\frac{3L}{2}\right) = (x-L)(x+2L)$$

69. $5R - 5r + r(R-r) = 5(R-r) + r(R-r) = (5+r)(R-r)$

71. $\sqrt{5+2\sqrt{6}} = \sqrt{2+2\sqrt{6}+3} = \sqrt{(\sqrt{2})^2+2\sqrt{6}+(\sqrt{3})^2} = \sqrt{(\sqrt{2}+\sqrt{3})^2} = \sqrt{2} + \sqrt{3}$

73. $\sqrt{71-16\sqrt{7}} = \sqrt{64-16\sqrt{7}+7} = \sqrt{8^2-16\sqrt{7}+(\sqrt{7})^2} = \sqrt{(8-\sqrt{7})^2} = 8 - \sqrt{7}$

75. $\sqrt{16+6\sqrt{7}} = \sqrt{9+6\sqrt{7}+7} = \sqrt{3^2+6\sqrt{7}+(\sqrt{7})^2} = \sqrt{(3+\sqrt{7})^2} = 3 + \sqrt{7}$

77. $\sqrt[3]{7+5\sqrt{2}} = \sqrt[3]{6+1+5\sqrt{2}} = \sqrt[3]{1+3\sqrt{2}+3(2)+2\sqrt{2}}$

$$= \sqrt[3]{1+3\sqrt{2}+3(\sqrt{2})^2+(\sqrt{2})^3} = \sqrt[3]{(1+\sqrt{2})^3} = 1 + \sqrt{2}$$

Section 1.6 Fractional Expressions

3. $\dfrac{3x}{4y^3} \div \dfrac{9x^4}{2y} = \dfrac{\cancel{3}x}{\cancel{4}y^3} \cdot \dfrac{\cancel{2}y}{\cancel{9}x^4} = \dfrac{xy}{6x^4y^3} = \dfrac{1}{6x^3y^2}$

5. $\dfrac{9b^2}{4c^3} \cdot \dfrac{24ac^2}{18b^5} \div \dfrac{c}{2a^3b} = \dfrac{\cancel{9}b^2}{\cancel{4}c^3} \cdot \dfrac{\overset{6}{\cancel{24}}ac^2}{\underset{2}{\cancel{18}}b^5} \cdot \dfrac{\cancel{2}a^3b}{c} = \dfrac{6a^4b^3c^2}{b^5c^4}$

$$= \dfrac{6a^4}{b^2c^2}$$

7. $\dfrac{\cancel{(x+3)}\cancel{(x-2)}}{\cancel{(x+3)}\cancel{(x-3)}} \cdot \dfrac{(x-3)^{\cancel{2}}}{2\cancel{(x-2)}} = \dfrac{x - 3}{2}$

9. $\dfrac{\overset{2}{\cancel{4}}xy^2}{(2x-3)\cancel{(2x+3)}} \cdot \dfrac{3\cancel{(2x+3)}}{\underset{5}{\cancel{10}}x^2y} = \dfrac{6xy^2}{5x^2y(2x-3)} = \dfrac{6y}{5x(2x-3)}$

11. $\dfrac{\cancel{x}\cancel{(x-y)}}{\cancel{x}\cancel{(x+y)}} \cdot \dfrac{x\cancel{(x+y)}}{y\cancel{(x-y)}} \cdot \dfrac{y(x-2)}{\cancel{x}(x-1)} = \dfrac{x(x-2)}{y(x-1)}$

13. $\dfrac{(2y+1)\cancel{(y+5)}}{\cancel{(2y+3)}(y+2)} \cdot \dfrac{\cancel{(2y+3)}\cancel{(y-1)}}{\cancel{(y+5)}\cancel{(y-1)}} = \dfrac{2y+1}{y+2}$

15. $\dfrac{(2x+3)\cancel{(x-1)}}{\cancel{(3x-2)}\cancel{(x+4)}} \cdot \dfrac{x\cancel{(3x-2)}}{\cancel{(x-1)}\cancel{(x+1)}} \cdot \dfrac{\cancel{(x+4)}\cancel{(x+1)}}{(x+6)} = \dfrac{x(2x+3)}{x+6}$

17. $\dfrac{s^2 - 2s - 3}{s - 2} \cdot \dfrac{(s+1)(s-2)}{(s-3)(s+3)} = \dfrac{\cancel{(s-3)}(s+1)}{\cancel{s-2}} \cdot \dfrac{(s+1)\cancel{(s-2)}}{\cancel{(s-3)}(s+3)} = \dfrac{(s+1)^2}{s+3}$

19. $\dfrac{a(x+y) + b(x+y)}{a(x-y) + b(x-y)} \cdot \dfrac{x - y}{x + y} = \dfrac{\cancel{(a+b)}\cancel{(x+y)}}{\cancel{(a+b)}\cancel{(x-y)}} \cdot \dfrac{\cancel{(x-y)}}{\cancel{(x+y)}} = 1$

21. $\dfrac{(2y+1)\cancel{(y-3)}}{\cancel{y-4}} \cdot \dfrac{\cancel{(y-4)}(y+4)(3-y)}{(y-3)^2} = \dfrac{(2y+1)(y+4)(-1)\cancel{(y-3)}}{\cancel{(y-3)}} = -(2y+1)(y+4)$

23. $\dfrac{\cancel{(a+200)}(a+100)}{(40-a)(40+a)} \cdot \dfrac{(a-40)(a+400)}{2\cancel{(a+200)}} = \dfrac{(a+100)\cancel{(a-40)}(a+400)}{(-1)\cancel{(a-40)}(a+40)(2)} = \dfrac{-(a+100)(a+400)}{2(a+40)}$

25. $\dfrac{2(ac)}{b(ac)} + \dfrac{1(ab)}{c(ab)} - \dfrac{3(bc)}{a(bc)} = \dfrac{2ac + ab - 3bc}{abc}$

27. $\dfrac{2}{3s} - \dfrac{1}{t} + \dfrac{1}{3st} = \dfrac{2(t)}{3st} - \dfrac{1(3s)}{3st} + \dfrac{1}{3st} = \dfrac{2t - 3s + 1}{3st}$

29. $\dfrac{3}{xy^2} + \dfrac{2(y^2)}{xy^2} - \dfrac{3(xy)}{xy^2} = \dfrac{3 + 2y^2 - 3xy}{xy^2}$

31. $\dfrac{3(x-3)}{(x+1)(x-3)} + \dfrac{2(x+1)}{(x+1)(x-3)} = \dfrac{3(x-3) + 2(x+1)}{(x+1)(x-3)} = \dfrac{3x - 9 + 2x + 2}{(x+1)(x-3)} = \dfrac{5x - 7}{(x+1)(x-3)}$

33. $\dfrac{x}{x - 5} + \dfrac{2}{x + 3} = \dfrac{x(x+3) + 2(x-5)}{(x-5)(x+3)} = \dfrac{x^2 + 3x + 2x - 10}{(x-5)(x+3)} = \dfrac{x^2 + 5x - 10}{(x-5)(x+3)}$

35. $\dfrac{3(x+2) - 2x - 1}{x(x+2)} = \dfrac{3x + 6 - 2x - 1}{x(x+2)} = \dfrac{x + 5}{x(x+2)}$

37. $\dfrac{2(2a+3) + 3a - 2}{a(2a+3)} = \dfrac{4a + 6 + 3a - 2}{a(2a+3)} = \dfrac{7a + 4}{a(2a+3)}$

39. $\dfrac{4(3x+4) - (9x-16)}{3x + 4} = \dfrac{12x + 16 - 9x + 16}{3x + 4} = \dfrac{3x + 32}{3x + 4}$

41. $\dfrac{2(p+1)^2 + 2p - p(p+1)}{p(p+1)^2} = \dfrac{\overset{2p^2+4p+2+2p-p^2-p}{2(p^2+2p+1) + 2p - p^2 - p}}{p(p+1)^2}$

$= \dfrac{2p^2 + 4p + 2 + p - p^2}{p(p+1)^2} = \dfrac{p^2 + 5p + 2}{p(p+1)^2}$

43. $\dfrac{(x+1) + (x-2) - (x+3)}{(x+1)(x-2)(x+3)} = \dfrac{x - 4}{(x+1)(x-2)(x+3)}$

45. $\dfrac{5x + 6}{(x+3)(x-2)} + \dfrac{x + 1}{(x-2)^2} - \dfrac{\frac{1}{\cancel{x}}}{\cancel{x^2}(x-2)}$

$$= \dfrac{(5x+6)(x-2) + (x+1)(x+3) - (x+3)(x-2)}{(x+3)(x-2)^2}$$

$$= \dfrac{(5x^2-4x-12) + (x^2+4x+3) - (x^2+x-6)}{(x+3)(x-2)^2} = \dfrac{5x^2 - x - 3}{(x-2)^2(x+3)}$$

47. $\dfrac{x + 3}{\sqrt{x+2}} + \sqrt{x-2} = \dfrac{x + 3 + (\sqrt{x-2})^2}{\sqrt{x-2}} = \dfrac{2x + 1}{\sqrt{x+2}} \cdot \dfrac{\sqrt{x+2}}{\sqrt{x-2}} = \dfrac{(2x+1)\sqrt{x-2}}{x - 2}$

49. $\dfrac{\sqrt{8}}{\sqrt{3} + \sqrt{5}} \cdot \dfrac{\sqrt{3} - \sqrt{5}}{\sqrt{3} - \sqrt{5}} = \dfrac{2\sqrt{2}(\sqrt{3}-\sqrt{5})}{3 - 5} = \dfrac{2\sqrt{2}(\sqrt{3}-\sqrt{5})}{-2}$

$$= \sqrt{2}(\sqrt{5}-\sqrt{3})$$

51. $\dfrac{\sqrt{x} - \sqrt{y}}{\sqrt{x} + \sqrt{y}} \cdot \dfrac{\sqrt{x} - \sqrt{y}}{\sqrt{x} - \sqrt{y}} = \dfrac{(\sqrt{x} - \sqrt{y})^2}{x - y} = \dfrac{x - 2\sqrt{xy} + y}{x - y}$

53. $\dfrac{\frac{x^2 - 1}{x}}{\frac{x - 1}{x}} = \dfrac{x^2 - 1}{\cancel{x}} \cdot \dfrac{\cancel{x}}{x - 1} = \dfrac{\cancel{(x-1)}(x+1)}{\cancel{x-1}} = x + 1$

55. $\dfrac{\frac{a^2 - b^2}{ab}}{\frac{a + b}{a}} = \dfrac{(a-b)\cancel{(a+b)}}{\cancel{ab}} \cdot \dfrac{\cancel{a}}{\cancel{a+b}} = \dfrac{a - b}{b}$

57. $\dfrac{1 - \frac{1}{x^2}}{\frac{1}{x} - 1} = \dfrac{\frac{x^2 - 1}{x^2}}{\frac{1 - x}{x}} = \dfrac{\cancel{(x-1)}(x+1)}{x^2} \cdot \dfrac{\cancel{x}}{\underset{-1}{-\cancel{(x-1)}}} = \dfrac{x + 1}{-x} = \dfrac{-(x+1)}{x}$

59. $\dfrac{\frac{c^2 + d^2}{cd}}{\frac{d + c}{cd}} = \dfrac{c^2 + d^2}{cd} \cdot \dfrac{cd}{c + d} = \dfrac{c^2 + d^2}{c + d}$

61. $\dfrac{\frac{2}{x - 1} - x - 2}{1 - x + \frac{2}{x + 2}} = \dfrac{\frac{2 - x(x-1) - 2(x-1)}{x - 1}}{\frac{x + 2 - x(x+2) + 2}{x + 2}} = \dfrac{\cancel{-x^2 - x + 4}}{x - 1} \cdot \dfrac{x + 2}{\cancel{-x^2 - x + 4}} = \dfrac{x + 2}{x - 1}$

63. $\dfrac{\dfrac{x(x-1)\,+\,(x-4)}{(x-4)(x-1)}}{\dfrac{x(x-3)\,+\,2(x-1)}{(x-1)(x-3)}} = \dfrac{x^2\,-\,4}{(x-4)(x-1)} \cdot \dfrac{(x-1)(x-3)}{x^2\,-\,x\,-\,2}$

$= \dfrac{(x+2)(x-2)(x-3)}{(x-4)(x-2)(x+1)} = \dfrac{(x+2)(x-3)}{(x-4)(x+1)}$

65. $\dfrac{\dfrac{x\,-\,y\,+\,y}{x\,-\,y}}{x\,-\,\dfrac{y}{\dfrac{x+y-x}{x+y}}} = \dfrac{\dfrac{x}{x\,-\,y}}{x\,-\,\cancel{y}\cdot\dfrac{(x+y)}{\cancel{y}}} = \dfrac{x}{x\,-\,y}\cdot\dfrac{1}{-y} = \dfrac{x}{-y(x-y)} = \dfrac{x}{y(y-x)}$

67. $\dfrac{4t^3\,+\,2(3t^2)\,-\,5(6)}{12t^3} = \dfrac{2(2t^3+3t^2-15)}{2(6t^3)} = \dfrac{2t^3\,+\,3t^2\,-\,15}{6t^3}$

69. $\dfrac{3}{s(s+1)} + \dfrac{2}{(s+3)(s+1)} = \dfrac{3(s+3)\,+\,2s}{s(s+3)(s+1)} = \dfrac{5s\,+\,9}{s(s+1)(s+3)}$

Section 1.7   Common Errors

5. $\frac{1}{2} + \frac{3}{4} = \frac{2\,+\,3}{4} = \frac{5}{4}$ 7. $\frac{x}{x} + \frac{2}{x} = 1 + \frac{2}{x}$ 9. $\left(\frac{1}{x}+\frac{1}{y}\right)^{-1} = \left(\frac{x+y}{xy}\right)^{-1} = \frac{xy}{x\,+\,y}$

Chapter 1   Review

15. $|5-9| = |-4| = 4$

17. $\left|-10-(-4)\right| = |-10+4| = |-6| = 6$

19. $-22 - |-39| = -22 - 39 = -61$ 21. $d = |5-14| = |-9| = 9$

23. $d = |-2-0| = |-2| = 2$ 25. $s = \left|-10-(-7)\right| = |-3| = 3$

27. $a - 2a - b + c = -a - b + c$

29. $5x - 6\big[2x-3x-7y-y\big] = 5x - 6\big[-x - 8y\big] = 5x + 6x + 48y = 11x - 48y$

33. $x(3x+7y) - 2y(3x+7y) = 3x^2 + 7xy - 6xy - 14y^2 = 3x^2 + xy - 14y^2$

35. $2y(x-3y+5) + 3(x-3y+5) = 2xy - 6y^2 + 10y + 3x - 9xy + 15$
    $= 2xy - 6y^2 + y + 3x + 15$

39. $2^3 + 3(2^2)3y + 3(2)(3y)^2 + (3y)^3 = 8 + 36y + 54y^2 + 27y^3$

59. $a(b-x) - 3(b-x) = (a-3)(b-x)$

61. $\dfrac{2x}{3xy} \cdot \dfrac{5x^2y^3}{4xy^4} \cdot \dfrac{3x^3y^2}{15x^4y} = \dfrac{x^6y^5}{6x^6y^6} = \dfrac{1}{6y}$

63. $\dfrac{(a+3)^2}{(a-b)(a+b)} \cdot \dfrac{a(a+b)}{b(a+3)} = \dfrac{a(a+3)}{b(a-b)}$

65. $\dfrac{(r-2)^2}{(r-3)(r+3)} \cdot \dfrac{(r-3)(r+1)}{(r-2)(r+2)} \cdot \dfrac{\cancel{r}(r+3)}{(r-2)} = \dfrac{\cancel{r}(r+1)}{r+2}$

67. $\dfrac{2a}{ab} - \dfrac{3a}{ab} + \dfrac{1}{ab} = \dfrac{2b - 3a + 1}{ab}$

69. $\dfrac{2x(x+3) + 3x - (x+3)}{x(x+3)} = \dfrac{2x^2 + 8x - 3}{x(x+3)}$

71. $\dfrac{a}{a-2} + \dfrac{1}{(a-5)^2} = \dfrac{a(a-5)^2 - (a-2)}{(a-2)(a-5)^2} = \dfrac{a^3 - 10a^2 + 25a - a + 2}{(a-2)(a-5)^2} = \dfrac{a^3 - 10a^2 + 24a + 2}{(a-2)(a-5)^2}$

73. $\dfrac{\dfrac{x - y - x - y}{(x+y)(x-y)}}{\dfrac{x - y + x + y}{(x+y)(x-y)}} = \dfrac{-2y}{\cancel{(x+y)(x-y)}} \cdot \dfrac{\cancel{(x+y)(x-y)}}{2x} = \dfrac{-y}{x}$

77. $\left(\dfrac{x}{y^2}+1\right)^{-1} = \left(\dfrac{x+y^2}{y^2}\right)^{-1} = \dfrac{y^2}{x + y^2}$

79. $\dfrac{x+\dfrac{1}{y}}{3+\dfrac{1}{x^2}+\dfrac{1}{y^2}} = \dfrac{\dfrac{xy+1}{y}}{\dfrac{3x^2y^2+y^2+x^2}{x^2y^2}} = \dfrac{xy+1}{y} \cdot \dfrac{x^2y^2}{3x^2y+y^2+x^2} = \dfrac{x^2y(xy+1)}{3x^2y^2+y^2+x^2}$

81. $\left(\dfrac{x^3y^{-7}}{2x^5y^{-2}}\right)^{-3} = \left(\dfrac{1}{2x^2y^5}\right)^{-3} = (2x^2y^5)^3 = 8x^6y^{15}$

83. $\dfrac{\dfrac{1}{x} + \dfrac{1}{x^2}}{\dfrac{1}{x} - \dfrac{1}{x^2}} = \dfrac{\dfrac{x^2 + x}{x^2}}{\dfrac{x^2 - x}{x^2}} = \dfrac{\cancel{x}(x+1)}{\cancel{x^2}} \cdot \dfrac{\cancel{x^2}}{\cancel{x}(x-1)} = \dfrac{x+1}{x+1}$

85. $\dfrac{\dfrac{1}{a^2}}{1 - \dfrac{1}{a}} - \dfrac{\dfrac{1}{a}}{1 - \dfrac{1}{a^2}} = \dfrac{\dfrac{1}{a^2}}{\dfrac{a-1}{a}} - \dfrac{\dfrac{1}{a}}{\dfrac{a^2-1}{a^2}}$

$= \dfrac{1}{a^2} \cdot \dfrac{a}{a-1} - \dfrac{1}{a} \cdot \dfrac{a^2}{a^2-1} = \dfrac{1}{a(a-1)} - \dfrac{a}{(a-1)(a+1)}$

$= \dfrac{a+1-a^2}{a(a-1)(a+1)} = \dfrac{1+a-a^2}{a(a-1)(a+1)}$

87. $\sqrt{16 \cdot 2} - \sqrt{9 \cdot 2} = 4\sqrt{2} - 3\sqrt{2} = \sqrt{2}$

89. $\sqrt{16 \cdot 3} + 5 \cdot 7 - 3\sqrt{3} = 4\sqrt{3} + 35 - 3\sqrt{3} = \sqrt{3} + 35$

91. $\dfrac{\sqrt{2}}{5+\sqrt{2}} \cdot \dfrac{5-\sqrt{2}}{5-\sqrt{2}} = \dfrac{5\sqrt{2}-2}{25-2} = \dfrac{5\sqrt{2}-2}{23} = \dfrac{\sqrt{2}(5-\sqrt{2})}{23}$

93. $\sqrt[5]{\sqrt{\sqrt[3]{9}}} = \left(\left(9^{1/3}\right)^{1/2}\right)^{1/5} = \left(9^{1/6}\right)^{1/5} = 9^{1/30} = (3^2)^{1/30} = 3^{1/15} = \sqrt[15]{3}$

95. $\sqrt{\dfrac{2}{c}+\dfrac{1}{d^2}} = \left(\dfrac{2d^2+c}{cd^2}\right)^{1/2} = \dfrac{1}{d}\left(\dfrac{2d^2+c}{c}\right)^{1/2} = \dfrac{1}{d}\left(\dfrac{c(2d^2+c)}{c^2}\right)^{1/2} = \dfrac{1}{cd}\left(2cd^2+c^2\right)^{1/2}$

$= \dfrac{1}{cd}\sqrt{2cd^2+c^2}$

97. $\dfrac{\sqrt{x}-\sqrt{y}}{\sqrt{x}+\sqrt{y}} \cdot \dfrac{\sqrt{x}-\sqrt{y}}{\sqrt{x}-\sqrt{y}} = \dfrac{(\sqrt{x}-\sqrt{y})^2}{x-y} = \dfrac{x-2\sqrt{xy}+y}{x-y}$

99. $\left[a^4(bc^3)^{-1}\right]^{1/3} = a^{4/3}(bc^3)^{-1/3} = \dfrac{\sqrt[3]{a^3 a}}{\sqrt[3]{bc^3}} = \dfrac{a \cdot \sqrt[3]{a}}{c \cdot \sqrt[3]{b}} \cdot \dfrac{\sqrt[3]{b^2}}{\sqrt[3]{b^2}}$

$= \dfrac{a \cdot \sqrt[3]{ab^2}}{bc}$

101. $a^{1/3}a^{1/2} + a^{1/3}b + a^{1/2}b^{1/2} + bb^{1/2} = a^{5/6} + a^{1/3}b + a^{1/2}b^{1/2} + b^{3/2}$

103. $\left(x^{1/2} - \dfrac{1}{x^{1/2}}\right)^{-1} = \left(\dfrac{x-1}{x^{1/2}}\right)^{-1} = \dfrac{x^{1/2}}{x-1}$

105. $4x^{-1} = \dfrac{4}{x}$

107. $\left(\dfrac{1}{a^{1/3}} - 2b^{1/3}\right)^2 = \left(\dfrac{1-2a^{1/3}b^{1/3}}{a^{1/3}}\right)^2 = \dfrac{1}{a^{2/3}}\left(1 - 4a^{1/3}b^{1/3} + 4a^{2/3}b^{2/3}\right)$

$= \dfrac{1}{a^{2/3}}\left(1 - 4(ab)^{1/3} + 4(ab)^{2/3}\right)$

109. $\left(ax^{1/2} + bx^{1/4} + c\right)\left(ax^{1/2} + bx^{1/4} + c\right)$

$= a^2 x + abx^{3/4} + acx^{1/2} + abx^{3/4} + b^2 x^{1/2} + bcx^{1/4} + cax^{1/2} + cbx^{1/4} + c^2$

$= a^2 x + b^2 x^{1/2} + c^2 + 2abx^{3/4} + 2acx^{1/2} + 2bcx^{1/4}$

111. $3 + \dfrac{5}{x+1} - \dfrac{x}{(x-3)(x+1)} = \dfrac{3(x^2-2x-3) + 5(x-3) - x}{(x-3)(x+1)} = \dfrac{3x^2 - 2x - 24}{(x-3)(x+1)}$

113. $F = (25-2t)(25+23t-2t^2)^{-1} = \dfrac{25-2t}{25+23t-2t^2} = \dfrac{\overset{1}{\cancel{25-2t}}}{\cancel{(25-2t)}(1+t)} = \dfrac{1}{1+t}$

Chapter 2 Equations and Inequalities

<u>Section 2.1   Linear Equations</u>

1. $2x+5 = 2$

$2x = -3$

$x = -\frac{3}{2}$

3. $6x-3-x = 4-x-3$

$5x-3 = -x+1$

$6x = 4$

$x = \frac{2}{3}$

5. $3x+5 = 4x-8$

$13 = x$

7. $\frac{4a}{3}-5a+2 = \frac{a}{2}-1$

$6\left(\frac{4a}{2}-5a+2\right) = \left(\frac{a}{2}-1\right)6$

$8a-30a+12 = 3a-6$

$-25a = -18$

$a = \frac{18}{25}$

9. $\frac{2x+1}{3}+16 = 3x$

$2x+1+48 = 9x$

$-7x = -49$

$x = 7$

11. $\frac{5}{c} = 10$

$5 = 10c$

$\frac{1}{2} = c$

13. $\frac{3x}{x+5}-4 = 0$

$3x-4(x+5) = 0$

$-x-20 = 0$

$x = -20$

15. $\frac{3}{6y+2} = \frac{4}{7y+3}$

$3(7y+3) = 4(6y+2)$

$21y+9 = 24y+8$

$1 = 3y$

$\frac{1}{3} = y$

17. $\frac{2}{x-2}-\frac{3}{x+5} = \frac{10}{(x-2)(x+5)}$

$2(x+5)-3(x-2) = 10$

$2x+10-3x+6 = 10$

$x = 6$

19.

$$\frac{5}{5m-11} = \frac{3}{m-5} - \frac{4}{2m-3}$$

$$5(m-5)(2m-3) = 3(5m-11)(2m-3) - 4(5m-11)(m-5)$$

$$5(2m^2-13m+15) = 3(10m^2-37m+33) - 4(5m^2-36m+55)$$

$$10m^2-65m+75 = 30m^2-111m+99 - 20m^2+144m-220$$

$$-65m+75 = 33m-121$$

$$-98m = -196$$

$$m = 2$$

21.

$$x^2+6x-7 = x^2+18x+81$$

$$-12x = 88$$

$$x = -\frac{88}{12} = -\frac{22}{3}$$

23.

$$2x^2+1 = 2x^2+5x-3$$

$$4 = 5x$$

$$x = \frac{4}{5}$$

25.

$$x^3+3x = 5+x^3$$

$$3x = 5$$

$$x = \frac{5}{3}$$

27.

$$\frac{5}{x+1} = \frac{3}{x+1} + \frac{2}{x-1}$$

$$5(x-1) = 3(x-1) + 2(x+1)$$

$$5x-5 = 3x-3+2x+2$$

$$-5 = -1$$

No solution

29.

$x \geq 0$	$x < 0$
$x+2 = -2$	$-x+2 = -2$
$x = -4$	$x = 4$
Impossible	Impossible

31.

$x \geq 0$	$x < 0$
$3x-1 = x$	$3x-1 = -x$
$2x = 1$	$4x = 1$
$x = \frac{1}{2}$	$x = \frac{1}{4}$
	Impossible

33.

$2x-3 \geq 0$	$2x-3 < 0$
$2x \geq 3$	$2x < 3$
$x \geq \frac{3}{2}$	$x < \frac{3}{2}$

$2x-3 = -1$	$-(2x-3) = -1$
$2x = 2$	$-2x+3 = -1$
$x = 1$	$-2x = -4$
Impossible	$x = 2$
	Impossible

35.

$x+5 \geq 0$	$x+5 < 0$
$x \geq -5$	$x < -5$
$(x+5)-2x = 1+x$	$-(x+5)-2x = 1+x$
$-2x = -4$	$-4x = 6$
$x = 2$	$x = -\frac{6}{4} = -\frac{3}{2}$
	Impossible

37.

$5x-1 \geq 0$	$5x-1 < 0$
$5x \geq 1$	$5x < 1$
$x \geq \frac{1}{5}$	$x < \frac{1}{5}$
$(5x-1)+2 = 3x+39$	$-(5x-1)+2 = 3x+39$
$2x = 38$	$-8x = 36$
$x = 19$	$x = -\frac{36}{8} = -\frac{9}{2}$

Section 2.2   Formulas and Applications of Linear Equations

21.
$$v = at+b$$
$$v-b = at$$
$$\frac{v-b}{a} = t$$

23.
$$A = \tfrac{1}{2}h(a+b)$$
$$\frac{2a}{h} = a+b$$
$$\frac{2a}{h} - a = b$$

25.
$$L = L_0\Big[1+\alpha(T_1-T_0)\Big]$$
$$L = L_0+\alpha L_0 T_1 - \alpha L_0 T_0$$
$$L - L_0+\alpha L_0 T_0 = \alpha L_0 T_1$$
$$\frac{L-L_0+\alpha L_0 T_0}{\alpha L_0} = T_1$$
$$T_0+\frac{L-L_0}{\alpha L_0} = T_1$$
$$T_0+\frac{1}{\alpha}\Big(\frac{L}{L_0}-1\Big) = T_1$$

27.
$$D = \frac{nd}{n+2}$$
$$D(n+2) = nd$$
$$Dn + 2D = nd$$
$$Dn - nd = -2D$$
$$n(D-d) = -2D$$
$$n = \frac{-2D}{D-d} = \frac{-2D}{-(d-D)}$$
$$n = \frac{2D}{d-D}$$

29.
$$\frac{1}{R} = \frac{1}{R_1} + \frac{1}{R_2}$$
$$R_1 R_2 = RR_2 + RR_1$$
$$R_1 R_2 = R(R_2 + R_1)$$
$$\frac{R_1 R_2}{R_1 + R_2} = R$$

31.
$$x + y = 20$$
$$x = 2 + y$$

$$(2+y) + y = 20$$
$$2y = 18$$
$$y = 9$$
$$x = 11$$

33.
$$x, \quad x+2, \quad x+4$$
$$x + (x+2) + (x+4) = 312$$
$$3x + 6 = 312$$
$$3x = 306$$
$$x = 102$$
$$102, \quad 104, \quad 106$$

35.

	rate	time	distance
jogger 1	$\frac{1 \text{ mi}}{7 \text{ min}}$	t	$\frac{t}{7}$
jogger 2	$\frac{1 \text{ mi}}{6 \text{ min}}$	t	$\frac{t}{6}$

$$d_1 = d_2 - \frac{1}{10}$$
$$\frac{t}{7} = \frac{t}{6} - \frac{1}{10}$$
$$60\,t = 70t - 42$$
$$10t = 42$$
$$t = 4.2$$
$$d_2 = \frac{4.2}{6} = 0.7 \text{ mile}$$

37. city $= x$
highway $= 500 - x$

$$\frac{x}{22} + \frac{500 - x}{28} = 20$$

$$28x + 22(500 - x) = 20(22)(28)$$

$$6x = 12{,}320 - 11{,}000 = 1320$$

$$x = \frac{1320}{6} = 220$$

39a. 75% wholesale price

$$x = 100 + 10(.75) = \$175$$

39b. 75% retail price

$$x = 100 + .75x$$

$$.25x = 100$$

$$x = \$400$$

41.

	rate	time	distance
With wind	$400 + x$	1	$(400 + x)1$
Against wind	$400 - x$	2	$(400 - x)2$

$$400 + x = (400 - x)2$$

$$3x = 400$$

$$x = \frac{400}{3} = 133.3$$

43.

	rate	distance
Fast	170	$x + 2.5$
Slow	165	x

$$t = \frac{d}{r}$$

$$\frac{x + 2.5}{170} = \frac{x}{165}$$

$$165(x + 2.5) = 170x$$

$$2.5(165) = 5x$$

$$82.5 = x$$

$$t = \frac{82.5}{165} = .5$$

45. touchdown $= t$ field goals $= f$

extra points $= e$ safeties $= s$

$$t = e = f = 2s$$

$$t + e + f + s = 44$$

$$6t + e + 3f + 2s = 44$$

$$6t + t + 3t + 2\left(\frac{t}{2}\right) = 44$$

$$11t = 44$$

$$t = 4, \ e = 4$$

$$f = 4, \ s = 2$$

47. Girl $= x$

Boy $= 20 - x$

$$100x = 150(20 - x)$$

$$100x = 3000 - 150x$$

$$250x = 3000$$

$$x = 12 \text{ feet from Girl}$$

49.

	rate	time	distance
car	55	$x + .5$	$55(x + .5)$
truck	45	x	$45x$
total			206

$$55(x + .5) + 45x = 206$$

$$100x + 27.5 = 206$$

$$x = 1.785$$

Section 2.3   Quadratic Equations

1. $x(x - 2) = 0$

$x = 0, \ 2$

3. $(x - 2)(x + 1) = 0$

$x = -1, \ 2$

5. $x^2 - x - 6 = 0$

$(x - 3)(x + 2) = 0$

$x = 3, \ -2$

7. $(z + 5)(z - 2) = 0$

$z = -5, \ 2$

9. $(2x - 3)(x + 4) = 0$

$x = \frac{3}{2}, \ -4$

11. $(3w + 1)(w - 2) = 0$

$w = -\frac{1}{3}, \ +2$

13. $2v(5v - 1) = 0$

$v = 0, \ \frac{1}{5}$

15. $t^2 - 3t - 10 = 0$

$(t - 5)(t + 2) = 0$

$t = 5, \ -2$

17. $35x^2 + 2x - 24 = 0$

    $(7x+6)(5x-4) = 0$

       $x = -\frac{6}{7}, \frac{4}{5}$

19. $x = \dfrac{-4 \pm \sqrt{(-4)^2 - 4(3)(-4)}}{2(3)}$

    $= \dfrac{-4 \pm \sqrt{64}}{6} = \dfrac{-4+8}{6}$ or $\dfrac{-4-8}{6} = \dfrac{2}{3}, -2$

21. $x = \dfrac{-2 \pm \sqrt{2^2 - 4(5)(-3)}}{2(5)} = \dfrac{-2 \pm \sqrt{64}}{10} = \dfrac{-2 \pm 8}{10} = -1, \dfrac{3}{5}$

23. $x = \dfrac{1 \pm \sqrt{(-1^2) - 4(2)(-2)}}{2(2)} = \dfrac{1 \pm \sqrt{17}}{4}$

25. $2x^2 - x - 4 = 0$

    $x = \dfrac{1 \pm \sqrt{(-1)^2 - 4(2)(-4)}}{2(2)} = \dfrac{1 \pm \sqrt{33}}{4}$

27. $x = \dfrac{5 \pm \sqrt{(-5)^2 - 4(1)(5)}}{2} = \dfrac{5 \pm \sqrt{5}}{2}$

29. $x = \dfrac{-12 \pm \sqrt{(12)^2 - 4(5)(6)}}{2(5)} = \dfrac{-12 \pm \sqrt{24}}{10} = \dfrac{-12 \pm 2\sqrt{6}}{10} = \dfrac{2(-6 \pm \sqrt{6})}{10} = \dfrac{-6 \pm \sqrt{6}}{5}$

31. $x = \dfrac{2 \pm \sqrt{(-2)^2 - 4(3)(2)}}{2(3)} = \dfrac{2 \pm \sqrt{-20}}{6} = \dfrac{2 \pm 2\sqrt{5}i}{6} = \dfrac{2(1 \pm \sqrt{5}i)}{6} = \dfrac{1 \pm \sqrt{5}i}{3}$

33. $x = \dfrac{2 \pm \sqrt{(-2)^2 - 4(2)(5)}}{2(2)} = \dfrac{2 \pm \sqrt{-36}}{4} = \dfrac{2 \pm 6i}{4} = \dfrac{2(1 \pm 3i)}{4} = \dfrac{1 \pm 3i}{2}$

35. $x = \dfrac{\pi \pm \sqrt{(-\pi)^2 - 4(3)(2)}}{2(3)} = \dfrac{\pi \pm \sqrt{\pi^2 - 24}}{6} = \dfrac{\pi \pm i\sqrt{24 - \pi^2}}{6}$

    $\simeq .524 \pm \dfrac{3.759}{6} = .524 \pm .627i$

37. $\sqrt{6}x^2 + \sqrt{6}x - 2.7 = 0$

    $x = \dfrac{-\sqrt{6} \pm \sqrt{6 - 4(\sqrt{6})(-2.7)}}{2\sqrt{6}} = -\dfrac{1}{2} \pm \dfrac{\sqrt{32.454}}{2\sqrt{6}} = -.5 \pm 1.163$

39. $x = \dfrac{.2 \pm \sqrt{(.2)^2 - 4(1.1)(.8)}}{2(1.1)} = \dfrac{.2 \pm \sqrt{-3.48}}{2.2} = .091 \pm .848i$

41. $x = \dfrac{3 \pm \sqrt{3^2 - 4(k)1}}{2k} = \dfrac{3 \pm \sqrt{9 - 4k}}{2k}$

$$\left(\dfrac{3 + \sqrt{9-4k}}{2k} \right) + \left(\dfrac{3 - \sqrt{9-4k}}{2k} \right) = \pi$$

$$3 + \sqrt{9-4k} + 3 - \sqrt{9-4k} = 2k\pi$$

$$6 = 2k\pi$$

$$\dfrac{3}{\pi} = k$$

43. $b^2 - 4ac = 1 - 4k8 \geq 0$

$$1 \geq 32k$$

$$\dfrac{1}{32} \geq k$$

45. $x = \dfrac{k \pm \sqrt{k^2 + 20}}{2}$

$$k + 20 = \text{an integer}$$

$$k = \pm 4$$

47. $x = \dfrac{k \pm \sqrt{k^2 - 4(k-1)}}{2} = \dfrac{k \pm \sqrt{k^2 - 4k + 4}}{2}$

$$\dfrac{k + \sqrt{k^2 - 4k + 4}}{2} = \dfrac{k - \sqrt{k^2 - 4k + 4}}{2}$$

$$2\sqrt{k^2 - 4k + 4} = 0$$

$$\sqrt{(k-2)^2} = 0$$

$$k = 2$$

49. $x = \dfrac{2k \pm \sqrt{(-2k)^2 - 4k(5)}}{2k} = \dfrac{2k \pm 2\sqrt{k^2 - 5k}}{2k} = \dfrac{2\left(k \pm \sqrt{k^2 - 5k}\right)}{2k} = \dfrac{k \pm \sqrt{k^2 - 5k}}{k}$

$$\dfrac{k + \sqrt{k^2 - 5k}}{k} = \dfrac{k + \sqrt{k^2 - 5k}}{k} + 3$$

$$2\sqrt{k^2 - 5k} = 3k$$

$$4(k^2 - 5k) = 9k^2$$

$$5k^2 + 20k = 0$$

$$5k(k+4) = 0$$

$$k = \cancel{0},\ -4$$

51. $x = \dfrac{-b \pm \sqrt{b^2 - 4ac}}{2a}$

$$\left(\frac{-b + \sqrt{b^2 - 4ac}}{2a}\right)\left(\frac{-b - \sqrt{b^2 - 4ac}}{2a}\right) = \frac{b^2 - \left(\sqrt{b^2 - 4ac}\right)^2}{4a} = \frac{b^2 - (b^2 - 4ac)}{4a} = \frac{4ac}{4a^2} = \frac{c}{a}$$

Section 2.4   Applied Problems That Lead to Quadratic Equations

1. $d^2 + 6d + 7 = 0$

$$d = \frac{-6 \pm \sqrt{6^2 - 4(1)7}}{2} = \frac{-6 \pm \sqrt{8}}{2} = \frac{2(-3 \pm \sqrt{2})}{2} = -3 \pm \sqrt{2}$$

3. $60 = 20x - x^2$

$x^2 - 20x + 60 = 0$

$$x = \frac{20 \pm \sqrt{20^2 - 4(1)(60)}}{2} = \frac{20 \pm 4\sqrt{10}}{2} = 10 \pm 2\sqrt{10}$$

5. $3 = \dfrac{x^2}{5(1-x)}$

$15(1-x) = x^2$

$x^2 + 15x - 15 = 0$

$$x = \frac{-15 \pm \sqrt{15^2 - 4(1)(-15)}}{2} = \frac{-15 \pm \sqrt{285}}{2}$$

7. $1100 = 100T - T^2$

$T^2 - 100T + 1100 = 0$

$$T = \frac{100 \pm \sqrt{100^2 - 4(1)(1100)}}{2} = \frac{100 \pm \sqrt{5600}}{2} = \frac{100 \pm 10\sqrt{56}}{2} = 87.4°, \, 12.6°$$

9. $x(5-x) + 7x = 10$

$5x - x^2 + 7x = 10$

$x^2 - 12x + 10 = 0$

$$x = \frac{12 \pm \sqrt{12^2 - 40}}{2} = \frac{12 \pm \sqrt{104}}{2} = 11.1, \, .9$$

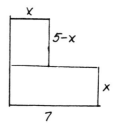

11.
$$x^2+(x+7)^2=13^2$$
$$x^2+x^2+14x+49=169$$
$$2x^2+14x-120=0$$
$$x^2+7x-60=0$$
$$(x+12)(x-5)=0$$
$$x=5,\ \cancel{-12}$$
sides: $x=5,\ x+7=12$

13.
$$w=x$$
$$l=x+50$$
$$x(x+50)=1875$$
$$x^2+50x-1875=0$$
$$(x+75)(x-25)=0$$
$$x=\cancel{-75},\ 25$$
$$w=25,\ l=75$$

15. Area mowed $=2(x-14)7+2\left(\frac{3}{4}x\right)7$

Area of interior $=\left(\frac{3}{4}x-14\right)(x-14)$

$$14(x-14)+\frac{21}{2}x=\left(\frac{3}{4}x-14\right)(x-14)$$
$$56(x-14)+42x=(3x-56)(x-14)$$
$$56x-784+42x=3x^2-98x+784$$
$$3x^2-196x+1568=0$$
$$x=\frac{196\pm\sqrt{196x^2-4(3)(1568)}}{6}=56,\ \cancel{9.3}$$
$$l=56,\ w=42$$

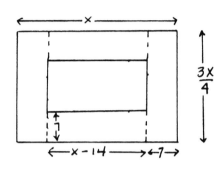

17.

	rate	distance	time
up	$6-x$	20	$20/(6-x)$
down	$6+x$	20	$20/(6+x)$

$$\frac{20}{6-x}+\frac{20}{6+x}=7.5$$
$$20(6+x)+20(6-x)=7.5(6-x)(6+x)$$
$$120+20x+120-20x=270-7.5x^2$$
$$7.5x^2-30=0$$
$$x^2-4=0$$
$$x=\pm2$$
rate of stream $=2$

19. hours in old week $= x$

 hours in new week $= x-5$

 rate in old week $= \frac{240}{x}$

 rate in new week $= \frac{245}{x-5}$

$$\frac{240}{x}+1 = \frac{245}{x-5}$$

$$240(x-5)+x(x-5) = 245x$$

$$240x-1200+x^2-5x = 245x$$

$$x^2-10x-1200 = 0$$

$$(x-40)(x+30) = 0$$

$$x = \cancel{-30},\ 40$$

21. numbers: $x,\ x+1,\ x+2$

$$x^2+(x+1)^2+(x+2)^2 = 434$$

$$x^2+x^2+2x+1+x^2+4x+4 = 434$$

$$3x^2+6x-429 = 0$$

$$x^2+2x-143 = 0$$

$$(x+13)(x-11) = 0$$

$$x = -13,\ 11$$

 numbers: $-13,\ -12,\ -11$ or $11,\ 12,\ 13$

23.

$$b = 6+h$$

$$A = \tfrac{1}{2}bh$$

$$20 = \tfrac{1}{2}(6+h)h$$

$$h^2+6h-40 = 0$$

$$(h-4)(h+10) = 0$$

$$h = 4,\ \cancel{-10}$$

$$h = 4,\ b = 6+4 = 10$$

25. large pipe $= x$

 small pipe $= x+4$

$$\frac{1}{x}+\frac{1}{x+4} = \frac{1}{5}$$

$$5(x+4)+5x = x(x+4)$$

$$5x+20+5x = x^2+4x$$

$$x^2-6x-20 = 0$$

$$x = \frac{6\pm\sqrt{(-6)^2+4(1)(-20)}}{2} = 8.4,\ \cancel{-2.4}$$

27. units $=x$

tens $=x-5$

$$2\Big(10(x-5)+x\Big) = 1+\Big(x^2+(x-5)^2\Big)$$
$$22x-100 = 1+x^2+x^2-10x+25$$
$$2x^2-32x+126 = 0$$
$$x^2-16x+63 = 0$$

$$x = \frac{16\pm\sqrt{16^2-4(1)(63)}}{2} = \frac{16\pm2}{2} = 7,\ 9$$

numbers $= 27$ or 49

Section 2.5   Equations in Quadratic Form

1. $\quad(x^2-4)(x^2+3) = 0$

$(x-2)(x+2)(x^2+3) = 0$

$x=2,\ -2$

3. $\quad(x^2-8)(x^2+2) = 0$

$(x-\sqrt{8})(x+\sqrt{8})(x^2+2) = 0$

$x=2\sqrt{2},\ -2\sqrt{2}$

5. $\quad(x^2-8)(x^2+1) = 0$

$x^2+8=0,\ x^2+1 = 0$

No real solutions

7. $2x^{-2}-5x^{-1}-3=0$

Let $y=x^{-1}$

$$2y^2-5y-3 = 0$$
$$(2y+1)(y-3) = 0$$

$y= -\tfrac{1}{2},\ 3$

$x= -2,\ \tfrac{1}{3}$

9. $3x^{-2}-4x^{-1}-4=0$

Let $y=x^{-1}$

$$3y^2-4y-4 = 0$$
$$(3y+2)(y-2) = 0$$

$y= -\tfrac{2}{3},\ 2$

$x= -\tfrac{3}{2},\ \tfrac{1}{2}$

11. $(x+2)^2-2(x+2)-8=0$

Let $y=x+2$

$$y^2-2y-8 = 0$$
$$(y-4)(y+2) = 0$$

$y=4,\ -2$

$x+2 = 4,\qquad x+2 = -2$

$x = 2\qquad\qquad x = -4$

13. $(x^2+2)+12(x^2+2)+11=0$

Let $y=x^2+2$

$$y^2+12y+11 = 0$$
$$(y+11)(y+1) = 0$$

$y = -11,\ -1$

$x^2+2 = -11,\qquad x^2+2 = -1$

$x^2 = -13\qquad\qquad x^2 = -3$

No real solutions

15. $(x-2)^{-2}+35=12(x-2)^{-1}$

Let $y=(x-2)^{-1}$

$$y^2-12y+35=0$$
$$(y-7)(y-5)=0$$
$$y=7,\ 5$$
$$\frac{1}{x-2}=7,\quad \frac{1}{x-2}=5$$

$$7x-14=1\qquad 5x-10=1$$
$$7x=15\qquad\quad 5x=11$$
$$x=\frac{15}{7}\qquad\quad x=\frac{11}{5}$$

17. $4(x^2+1)^2-7(x^2+1)-2=0$

Let $y=x^2+1$

$$4y^2-7y-2=0$$
$$(4y+1)(y-2)=0$$
$$y=-\frac{1}{4},\ 2$$

$$x^2+1=-\frac{1}{4}\qquad x^2+1=2$$
$$x^2=-\frac{5}{4}\qquad\quad x^2-1=0$$

No real solution $\qquad x=-1,\ 1$

19. $\frac{3}{x-1}-\frac{2}{x+3}=\frac{1}{2}$

$$3(2)(x+3)-2(2)(x-1)=(x-1)(x+3)$$
$$6x+18-4x+4=x^2+2x-3$$
$$x^2-25=0$$
$$(x-5)(x+5)=0$$
$$x=+5,\ -5$$

21. $\frac{1}{2}-\frac{1}{x-1}=\frac{6}{(x-1)(x+1)}$

$$(x-1)(x+1)-2(x+1)=6(2)$$
$$x^2-1-2x-2=12$$
$$x^2-2x-15=0$$
$$(x-5)(x+3)=0$$
$$x=-3,\ 5$$

23. $\frac{2}{x-1}+\frac{5}{x-7}=\frac{4}{3x-1}$

$$2(x-7)(3x-1)+5(x-1)(3x-1)=4(x-1)(x-7)$$
$$6x^2-44x+14+15x^2-20x+5=4x^2-32x+28$$
$$17x^2-32x-9=0$$
$$x=\frac{32\pm\sqrt{32^2-4(17)(-9)}}{2(17)}=\frac{32\pm\sqrt{1636}}{34}$$

25. $\dfrac{x}{(x-2)(x+2)}+\dfrac{1}{x-2}=\dfrac{x+5}{(x-2)(x+1)}$

$\qquad x(x-1)+(x+2)(x-1)=(x+5)(x+2)$

$\qquad\quad x^2-x+x^2+x-2=x^2+7x+10$

$\qquad\qquad\quad x^2-7x-12=0$

$\quad x=\dfrac{7\pm\sqrt{(-7)^2-4(1)(-12)}}{2}=\dfrac{7\pm\sqrt{97}}{2}$

27. $\dfrac{x^2-7}{x^2+4}+\dfrac{3}{x^2-1}=\dfrac{x^2+3}{x^2}$

$\qquad (x^2-7)x^2(x^2-1)+3(x^2+4)x^2=(x^2+3)(x^2+4)(x^2-1)$

$\qquad x^6-7x^4-x^4+7x^2+3x^4+12x^2=x^6+7x^4+12x^2-x^4-7x^2-12$

$\qquad\qquad\qquad 11x^4-14x^2-12=0$

$\quad$ Let $y=x^2$

$\quad 11y^2-14y-12=0$

$\qquad y=\dfrac{14\pm\sqrt{(-14)^2-4(11)(-12)}}{22}=\dfrac{14\pm\sqrt{724}}{22}=1.859,\ \cancel{-.0587}$

$\qquad x^2=1.859,\qquad x=\pm 1.3636$

29. $5\sqrt{x-6}=x$

$\quad 25(x-6)=x^2$

$\qquad x^2-25x+150=0$

$\qquad (x-10)(x-15)=0$

$\qquad x=10,\ 15$

$\qquad 5\sqrt{10-6}=10\ \checkmark$

$\qquad 5\sqrt{15-6}=15\ \checkmark$

31. $\sqrt{2y+3}=y$

$\quad 2y+3=y^2$

$\qquad y^2-2y-3=0$

$\qquad (y-3)(y+1)=0$

$\qquad y=\cancel{-1},\ 3$

$\qquad \sqrt{-2+3}=-1,\ \text{No}$

$\qquad \sqrt{6+3}=3,\ \checkmark$

33. $\sqrt{x^2-16}=3$

$\qquad x^2-16=9$

$\qquad x^2-25=0$

$\qquad x=5,\ -5$

35. $\sqrt{z^2+5z+2}=4$

$\qquad z^2+5z+2=16$

$\qquad z^2+5z-14=0$

$\qquad (z+7)(z-2)=0$

$\qquad\quad z=2,\ -7$

$\qquad \sqrt{4+10+2}=4\ \checkmark$

$\qquad \sqrt{49-35+2}=4\ \checkmark$

37.
$$y+1 = \sqrt{3y+7}$$
$$y^2+2y+1 = 3y+7$$
$$y^2-y-6 = 0$$
$$(y-3)(y+2) = 0$$

$$y = 3, \cancel{-2}$$

$$3+1 = \sqrt{9+7} \checkmark$$
$$-2+1 = \sqrt{-6+1}, \text{ No}$$

39.
$$2x-1 = \sqrt{2x+5}$$
$$4x^2-4x+1 = 2x+5$$
$$4x^2-6x-4 = 0$$
$$2x^2-3x-2 = 0$$
$$(2x+1)(x-2) = 0$$

$$x = \cancel{-\tfrac{1}{2}}, 2$$

$$-1-1 = \sqrt{-1+5}, \text{ No}$$
$$4-1 = \sqrt{4+5} \checkmark$$

41.
$$\sqrt{x+5} = \sqrt{x}-1$$
$$x+5 = x-2\sqrt{x}+1$$
$$4 = -2\sqrt{x}$$
$$16 = 4x$$
$$x = 4$$

$$\sqrt{4+5} = \sqrt{4}-1 \qquad \text{No solution}$$

43.
$$\sqrt{3y-5}-\sqrt{y+7} = 2$$
$$\sqrt{3y-5} = 2+\sqrt{y+7}$$
$$3y-5 = 4+4\sqrt{y+7}+y+7$$
$$y-8 = 2\sqrt{y+7}$$
$$y^2-16y+64 = 4(y+7)$$
$$y^2-20y+36 = 0$$
$$(y-18)(y-2) = 0$$
$$y = \cancel{2}, 18$$

$$\sqrt{6-5}-\sqrt{2+7} = 2, \text{ No}$$
$$\sqrt{54-5}-\sqrt{18+7} = 2 \checkmark$$

45.
$$\sqrt{5z+4} = \sqrt{z}+2$$
$$5z+4 = z+4\sqrt{z}+4$$
$$4z = 4\sqrt{z}$$
$$z^2 = z$$
$$z(z-1) = 0$$
$$z = 0, 1$$
$$\sqrt{0+4} = \sqrt{0}+2 \checkmark$$
$$\sqrt{5+4} = \sqrt{1}+2 \checkmark$$

47.

$$b^2+12^2 = c^2$$
$$b = \sqrt{c^2-144}$$
$$12^2+(b+4)^2 = (c+2)^2$$
$$144+b^2+8b+16 = c^2+4c+4$$
$$144+(c^2-144)+8\sqrt{c^2-144}+16 = c^2+4c+4$$
$$8\sqrt{c^2-144} = 4c-12$$
$$2\sqrt{c^2-144} = c-3$$
$$4(c^2-144) = c^2-6c+9$$
$$3c^2+6c-585 = 0$$
$$c^2+2c-195 = 0$$
$$(c+15)(c-13) = 0$$

$c= -\cancel{15},\ 13$

$a= 12, \quad b= 5, \quad c= 13$

$p= 12+13+5= 30$

49.

$$S = \pi r\sqrt{r^2+h^2}$$
$$S^2 = \pi^2 r^2(r^2+h^2)$$
$$S^2-\pi^2 r^4 = \pi^2 r^2 h^2$$
$$\frac{S^2-\pi^2 r^4}{\pi^2 r^2} = h^2$$
$$\frac{1}{\pi r}\sqrt{S^2-\pi^2 r^4} = h$$

51.

$$5 = 3+\sqrt{t+2}$$
$$2 = \sqrt{t+2}$$
$$4 = t+2$$
$$t = 2$$
$$5 = 3+\sqrt{4}\ \checkmark$$

Section 2.6   Linear Inequalities; Intervals

---

17. $3x+1 > x-4$

$2x > -5$

$x > -\frac{5}{2}$

19. $x+1 \geq \sqrt{2}-x\sqrt{3}$

$(1+\sqrt{3})x \geq \sqrt{2}-1$

$x \geq \frac{\sqrt{2}-1}{\sqrt{3}+1}$

21. $\frac{2x}{3}+5 < \frac{7x}{5}+\frac{1}{3}$

$10x+75 < 21x+5$

$70 < 11x$

$\frac{70}{11} < x$

23. $(x-3)(x+4) > (x-5)(x+5)$

$x^2+x-12 > x^2-25$

$x-12 > -25$

$x > -13$

25. $x^2 - 1 \geq (x+3)^2$

$x^2 - 1 \geq x^2 + 6x + 9$

$-10 \geq 6x$

$-\dfrac{5}{3} \geq x$

27. $x(x-1) > x(x-3)$

$x^2 - x > x^2 - 3x$

$2x > 0$

$x > 0$

29. $x^2 - 5x + 4 < x(x-2)$

$x^2 - 5x + 4 < x^2 - 2x$

$-3x < -4$

$x < \dfrac{4}{3}$

31. $x + 2 < 2x + 5 \leq 6x + 7$

$x + 2 < 2x + 5$ and $2x + 5 \leq 6x + 7$

$-3 < x$ $-2 \leq 4x$

                                $-\dfrac{1}{2} \leq x$

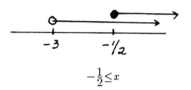

$-\dfrac{1}{2} \leq x$

33. $x < 2x + 3 < 3x - 5$

$x < 2x + 3$ and $2x + 3 < 3x - 5$

$-3 < x$ $8 < x$

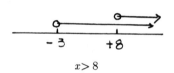

$x > 8$

35. $x^2 + 1 < (x-3)(x-4) \leq (x+3)(x+5)$

$x^2 + 1 < x^2 - 7x + 12 \leq x^2 + 8x + 15$

$1 < -7x + 12$ and $-7x + 12 \leq 8x + 15$

$7x < 11$ $-15x \leq 3$

$x < \dfrac{11}{7}$ $x \geq -\dfrac{1}{5}$

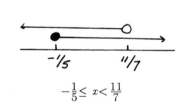

$-\dfrac{1}{5} \leq x < \dfrac{11}{7}$

41. $-3 < x - 2 < 3$

$-3 < x - 2$ and $x - 2 < 3$

$-1 < x$ $x < 5$

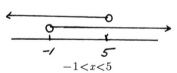

$-1 < x < 5$

43. $-5 < x + 2 < 5$

$-5 < x + 2$ and $x + 2 < 5$

$-7 < x$ $x < 3$

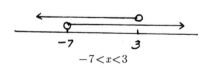

$-7 < x < 3$

45. $x - 5 > 1$ or $x - 5 < -1$

$x > 6$ $x < 4$

$(6, \infty) \cup (-\infty, 4)$

47. $-5 \le \frac{1}{2}x - 3 \le 5$

$$-10 \le x - 6 \quad \text{and} \quad x - 6 \le 10$$
$$-4 \le x \qquad\qquad x \le 16$$

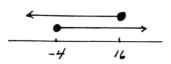

49. $-2 \le 2x - 3 \le 2$

$$-2 \le 2x - 3 \quad \text{and} \quad 2x - 3 \le 2$$
$$1 \le 2x \qquad\qquad 2x \le 5$$
$$\frac{1}{2} \le x \qquad\qquad x \le \frac{5}{2}$$

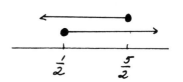

55. $2 < x < 6$

$$m = \frac{6+2}{2} = 4$$

$$hl = \frac{|2-6|}{2} = 2$$

$$|x - 4| < 2$$

57. $-10 < x < 16$

$$m = \frac{-10+16}{2} = 13$$

$$hl = \frac{|-10-16|}{2} = 13$$

$$|x - 3| < 13$$

61. $2 \le x \le 7$

$$m = \frac{2+7}{2} = \frac{9}{2}$$

$$hl = \frac{|2-7|}{2} = \frac{5}{2}$$

$$\left| x - \frac{9}{2} \right| \le \frac{5}{2}$$

$$|2x - 9| \le 5$$

65. $|M - 10| < .01$

$$-.01 < M - 10 < .01$$
$$9.99 < M < 10.01$$

1. $x^2 + 2x \geq 3$

$x^2 + 2x - 3 \geq 0$

$(x+3)(x-1) \geq 0$

| $x + 3$ | $-$ | $+$ | $+$ |
| $x - 2$ | $-$ | $-$ | $+$ |

$\begin{array}{ccc} & \quad & \\ \hline & + & + \\ -3 & & 1 \end{array}$

$x \leq -3$ or $x \geq 1$

3. $5x < 3x^2 - 2$

$3x^2 - 5x - 2 > 0$

$(3x+1)(x-1) > 0$

| $3x + 1$ | $-$ | $+$ | $+$ |
| $x - 2$ | $-$ | $-$ | $+$ |

$\begin{array}{ccc} & & \\ \hline -1/3 & & 2 \end{array}$

$x < -\dfrac{1}{3}$ or $x > 2$

5. $x(6-x) \geq 3x^2 + 2x - 15$

$0 \geq 4x^2 - 4x - 15$

$0 \geq (2x+3)(2x-5)$

| $2x + 3$ | $-$ | $+$ | $+$ |
| $2x - 5$ | $-$ | $-$ | $+$ |

$\begin{array}{ccc} & & \\ \hline -\frac{3}{2} & & \frac{5}{2} \end{array}$

$-\dfrac{3}{2} \leq x \leq \dfrac{5}{2}$

7. $x - 1 < x^2 + 1$

$0 < x^2 - x + 2$

$x < \dfrac{1 \pm \sqrt{1 - 8}}{2}$

All x

9. $x + 4 < x^2 - 4$

$0 < x^2 - x - 8$

$0 < \left(x - \dfrac{\left(1 + \sqrt{33}\right)}{2}\right)\left(x + \dfrac{\left(1 - \sqrt{33}\right)}{2}\right)$

| $x - \dfrac{(1+\sqrt{33})}{2}$ | $-$ | $-$ | $+$ |
| $x - \dfrac{(1-\sqrt{33})}{2}$ | $-$ | $+$ | $+$ |

$\begin{array}{ccc} & & \\ \hline \frac{1-\sqrt{33}}{2} & & \frac{1+\sqrt{33}}{2} \end{array}$

$x < \dfrac{1 - \sqrt{33}}{2}$ or $x > \dfrac{1 + \sqrt{33}}{2}$

11. $\dfrac{(x+4)(x+5)}{x+1}$

| $x+4$ | $-$ | $-$ | $+$ | $+$ |

| $x+5$ | $-$ | $+$ | $+$ | $+$ |

| $x+1$ | $-$ | $-$ | $-$ | $+$ |

$$-5 \qquad -4 \qquad -1$$

$x < -5$ or $-4 < x < -1$

13. $\dfrac{x+1}{2x+3} - 1 \leq 0$

$\dfrac{x+1-(2x+3)}{2x+3} \leq 0$

$\dfrac{-(x+2)}{2x+3} \leq 0$

| $-(x+2)$ | $+$ | $-$ | $-$ |

| $2x+3$ | $-$ | $-$ | $+$ |

$$-2 \qquad -\tfrac{3}{2}$$

$x \leq -2$ or $x > -\dfrac{3}{2}$

15. $\dfrac{(x-1)(x+1)(x^2+1)}{(x-2)(x+2)} \leq 0$

| $x-1$ | $-$ | $-$ | $-$ | $+$ | $+$ |

| $x+1$ | $-$ | $-$ | $+$ | $+$ | $+$ |

| x^2+1 | $+$ | $+$ | $+$ | $+$ | $+$ |

| $x-2$ | $-$ | $-$ | $-$ | $-$ | $+$ |

| $x+2$ | $-$ | $+$ | $+$ | $+$ | $+$ |

$$-2 \qquad -1 \qquad 1 \qquad 2$$

$-2 < x \leq -1$ or $1 \leq x < 2$

17. $\dfrac{x^2+1}{x^2+4} - 1 \leq 0$

$\dfrac{x^2+1-(x^2+4)}{x^2+4} \leq 0$

$\dfrac{-3}{x^2+4} \leq 0$

All x

19. $\dfrac{x-2}{x^2+x-1} \geq 0$

$$\dfrac{x-2}{\left(x-\dfrac{(-1+\sqrt{5})}{2}\right)\left(x-\dfrac{(-1-\sqrt{5})}{2}\right)} \geq 0$$

$x-2$	$-$	$-$	$-$	$+$
$x-\dfrac{\left(-1+\sqrt{5}\right)}{2}$	$-$	$-$	$+$	$+$
$x-\dfrac{\left(-1-\sqrt{5}\right)}{2}$	$-$	$+$	$+$	$+$

$$\begin{array}{c|c|c|c} & & & \\ \hline & \frac{-1-\sqrt{5}}{2} & \frac{-1+\sqrt{5}}{2} & 2 \end{array}$$

$$\dfrac{-1-\sqrt{5}}{2} < x < \dfrac{-1+\sqrt{5}}{2} \quad \text{or} \quad x \geq 2$$

21. $\dfrac{x+2}{x^2-3} \leq 1$

$$\dfrac{x+2-(x^2-3)}{x^2-3} \leq 0$$

$$\dfrac{-x^2+x+5}{x^2-3} \leq 0$$

$$\dfrac{x^2-x-5}{x^2-3} \geq 0$$

$$\dfrac{(x+1.79)(x-2.79)}{(x-\sqrt{3})(x+\sqrt{3})} \geq 0$$

$x+1.79$	$-$	$+$	$+$	$+$	$+$
$x+\sqrt{3}$	$-$	$-$	$+$	$+$	$+$
$x-\sqrt{3}$	$-$	$-$	$-$	$+$	$+$
$x-2.79$	$-$	$-$	$-$	$-$	$+$

$$\begin{array}{c|c|c|c|c} & -1.79 & -\sqrt{3} & \sqrt{3} & 2.79 \end{array}$$

$$x \leq -1.79 \quad \text{or} \quad -\sqrt{3} < x < \sqrt{3} \quad \text{or} \quad x \geq 2.79$$

23. $4x^2 - kx + 3 = 0$

$$k^2 - 4(4)(3) \geq 0$$
$$k^2 - 48 \geq 0$$
$$(k - \sqrt{48})(k + \sqrt{48}) \geq 0$$

| $k - \sqrt{48}$ | $-$ | $--$ | $+$ |
| $k + \sqrt{48}$ | $-$ | $+$ | $+$ |

$$-\sqrt{48} \qquad \sqrt{48}$$

$$k \leq -\sqrt{48} \quad \text{or} \quad k \geq \sqrt{48}$$

25. $15t - 3t^2 > 0$

$$3t(5 - t) > 0$$

| $3t$ | $-$ | $+$ | $+$ |
| $5 - t$ | $+$ | $+$ | $-$ |

$$0 \qquad 5$$

$$0 < t < 5$$

27. $500 + 600\,T - 20\,T^2 > 4500$

$$20\,T^2 - 600\,T + 4500 < 0$$
$$T^2 - 30\,T + 200 < 0$$
$$(T - 10)(T - 20) < 0$$

| $T - 10$ | $-$ | $+$ | $+$ |
| $T - 20$ | $-$ | $-$ | $+$ |

$$10 \qquad 20$$

$$10 < T < 20$$

Chapter 2   Review

1. $2x - 3(x - 5) = 12$

$$2x - 3x + 15 = 12$$
$$x = 3$$

3. $\frac{3x}{2} + 2(3x - 5) = \frac{x}{3}$

$$9x + 12(3x - 5) = 2x$$
$$43x = 60$$
$$x = \frac{60}{43}$$

5. $x - 3a(6 - 7x) = 19\,ax$

$$x - 18a + 21ax = 19ax$$
$$2ax + x = 18a$$
$$x(2a + 1) = 18a$$
$$x = \frac{18a}{2a + 1}$$

7. $\frac{2}{x} + \frac{3}{2x} = 5$

$$4 + 3 = 10x$$
$$\frac{7}{10} = x$$

9. $\dfrac{2}{2y-3} = \dfrac{5}{y+4}$

   $2(y+4) = 5(2y-3)$

   $23 = 8y$

   $\dfrac{23}{8} = y$

11. $\dfrac{4}{t-5} + 3 - \dfrac{3t}{t+4} = 0$

   $4(t+4) + 3(t+4)(t-5) - 3t(t-5) = 0$

   $4t + 16 + 3t - 3t - 60 - 3t^2 + 15t = 0$

   $16t = 44$

   $t = \dfrac{11}{4}$

13. $|x| + 3 = 7$

   $|x| = 4$

   $x = \pm 4$

15. $|x-4| = 10$

$x \geq 4$	$x < 4$
$x - 4 = 10$	$-(x-4) = 10$
$x = 14$	$x = -6$

17. $2y + |y-7| = 5$

$y - 7 > 0,\ y > 7$	$y - 7 < 0,\ y < 7$
$2y + y - 7 = 5$	$2y - (y-7) = 5$
$3y = 12$	$y + 7 = 5$
$y = 4$	$y = -2$
Impossible	

19. $3x^2 - 7x = 0$

   $x(3x - 7) = 0$

   $x = 0,\ \dfrac{7}{3}$

21. $(y - \sqrt{10})(y + \sqrt{10}) = 0$

   $y = \pm 10$

23. $7a = 2a^2 + 3$

   $2a^2 - 7a + 3 = 0$

   $(2a - 1)(a - 3) = 0$

   $a = \dfrac{1}{2},\ 3$

25. $z^2 + 6z + 5 = 0$

   $(z + 5)(z + 1) = 0$

   $z = -5,\ -1$

27. $y = \dfrac{4 \pm \sqrt{(-4)^2 - 4(4)(-11)}}{2(4)} = \dfrac{4 \pm 8\sqrt{3}}{8} = \dfrac{1 \pm 2\sqrt{3}}{2}$

29. $x = \dfrac{12 \pm \sqrt{(-12)^2 - 4(2)(-15)}}{4} = \dfrac{12 \pm \sqrt{264}}{4} = \dfrac{6 \pm \sqrt{66}}{2}$

31. $15y^2 + 14y - 8 = 0$

$$y = \frac{-14 \pm \sqrt{14^2 - 4(15)(-8)}}{2(15)} = \frac{-14 \pm \sqrt{676}}{30} = -\frac{4}{3}, \frac{2}{5}$$

33. $t = \dfrac{4 \pm \sqrt{(-4)^2 - 4(1)(8)}}{2} = \dfrac{4 \pm \sqrt{-16}}{2} = 2 \pm 2i$

35. $x = \dfrac{-7 \pm \sqrt{(7)^2 - 4(3)(-2)}}{2(3)} = \dfrac{-7 \pm \sqrt{73}}{6}$

37. $h^4 - 7h^2 + 12 = 0$

Let $y = h^2$

$$y^2 - 7y + 12 = 0$$
$$(y-3)(y-4) = 0$$

$$y = 3, 4$$
$$h^2 = 3, 4$$
$$h = \pm\sqrt{3}, \pm 2$$

39. $3(x-1)^2 - 10(x-1) - 8 = 0$

$$y = x - 1$$
$$3y^2 - 10y - 8 = 0$$
$$(3y+2)(y-4) = 0$$

$$y = -\frac{2}{3}, 4$$

$$x - 1 = -\frac{2}{3}, \qquad x - 1 = 4$$
$$x = \frac{1}{3}, \qquad\qquad x = 5$$

41. $\frac{1}{t+1} + \frac{1}{t} = \frac{9}{20}$

$$20t + 20(t+1) = 9t(t+1)$$
$$9t^2 - 31t - 20 = 0$$
$$(9t+5)(t-4) = 0$$

$$t = -\frac{5}{9}, 4$$

43. $\sqrt{x+12} - \sqrt{x} = 3$

$$\sqrt{x+12} = \sqrt{x} + 3$$
$$x + 12 = x + 6\sqrt{x} + 9$$
$$3 = 6\sqrt{x}$$
$$\frac{1}{2} = \sqrt{x}$$
$$\frac{1}{4} = x$$

45. $\sqrt{7t+23}-\sqrt{3t+7}=2$

$\sqrt{7t+23}=2+\sqrt{3t+7}$

$7t+23=4+4\sqrt{3t+7}+3t+7$

$4t+12=4\sqrt{3t+7}$

$t+3=\sqrt{3t+7}$

$t^2+6t+9=3t+7$

$t^2+3t+2=0$

$(t+2)(t+1)=0$

$t=-1,\ -2$

$\sqrt{-7+23}-\sqrt{-3+4}=2$ ✓

$\sqrt{-14+23}-\sqrt{-6+7}=2$ ✓

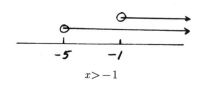

47. $x-4>2x+3$

$-7>x$

49. $3x+5\le 2-3x$

$6x\le -3$

$x\le -\frac{1}{2}$

51. $x(x-5)<(x+1)(x-2)$

$x^2-5x<x^2-x-2$

$2<4x$

$\frac{1}{2}<x$

53. $3\le 2x-5<8$

$3\le 2x-5\qquad 2x-5<8$

$8\le 2x\qquad\qquad 2x<13$

$4\le x\qquad\qquad x<\frac{13}{2}$

$4\le x<\frac{13}{2}$

55. $x-1<2x<3x+5$

$x-1<2x\quad$ and $\quad 2x<3+5$

$-1<x\qquad\qquad -5<x$

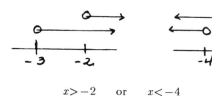

$x>-1$

57. $x+3\ge 0,\ x\ge -3\quad$ or $\quad x+3<0,\ x<-3$

$x+3>1\qquad\qquad\qquad -(x+3)>1$

$x>-2\qquad\qquad\qquad x+3<-1$

$x<-4$

$x>-2\quad$ or $\quad x<-4$

59. $-2 < x < 5$

$$m = \frac{-2+5}{2} = \frac{3}{2}$$

$$hl = \frac{|-2-5|}{2} = \frac{7}{2}$$

$$\left| x - \frac{3}{2} \right| < \frac{7}{2}$$

$$|2x - 3| < 7$$

61. $x \le -1 \quad$ or $\quad x \ge 3$

$$m = \frac{-1+3}{2} = 1$$

$$hl = \frac{|-1-3|}{2} = 2$$

$$|x - 1| \ge 2$$

63. $x^2 + 5x - 6 \le 0$

$(x+6)(x-1) \le 0$

$x+6$	$-$	$+$	$+$
$x-1$	$-$	$-$	$+$

$$\underrule{\quad -6 \qquad 1 \quad}$$

$-6 \le x \le 1$

65. $x + 2 > x^2 + 2x$

$0 > x^2 + x - 2$

$0 > (x+2)(x-1)$

$x+2$	$-$	$+$	$+$
$x-1$	$-$	$-$	$+$

$$\underrule{\quad -2 \qquad 1 \quad}$$

$-2 < x < 1$

67. $\dfrac{x-5}{(x-3)(x-5)} \ge 0$

$$\dfrac{1}{x-3} \ge 0$$

$x-3$	$-$	$+$

$$\underrule{\quad 3 \quad}$$

$x > 3, \; x \ne 5$

69. $P_1 - P_2 = \frac{1}{2}\rho\left(v_2{}^2 - v_1{}^2\right)$

$$\frac{2(P_1 - P_2)}{v_2{}^2 - v_1{}^2} = \rho$$

71. $E = c^2(m - m_0)$

$E = c^2 m - c^2 m_0$

$c^2 m_0 = c^2 m - E$

$$m_0 = \frac{c^2 m - E}{c^2}$$

73. $x + 3x = 787$

$4x = 787$

$x = 196.75$

$3x = 590.25$

75.

	amt.	concentration
S_1	x	.06
S_2	$10-x$	.02
$S_1 + S_2$	10	.05

$.06x + .02(10 - x) = .05(10)$

$4x = 30 \qquad x = 7.5 \text{ of } 6\% \qquad 10 - x = 2.5 \text{ of } 2\%$

77.

$$v = 5t - 16t^2$$

$$-100 = 5t - 16t^2$$

$$16t^2 - 5t - 100 = 0$$

$$t = \frac{5 \pm \sqrt{25 - 4(16)(-100)}}{32}$$

$$t = 2.66,\ -2.35$$

79.

$$\frac{1}{p} + \frac{1}{p-3} = \frac{1}{3}$$

$$3(p-3) + 3p = p(p-3)$$

$$p^2 - 9p + 9 = 0$$

$$p = \frac{9 \pm \sqrt{9^2 - 4(1)(9)}}{2}$$

$$= \frac{9 \pm 3\sqrt{5}}{2}$$

81.

$$m = m_0\left(1 - \frac{v^2}{c^2}\right)^{-\frac{1}{2}}$$

$$\frac{m}{m_0} = \left(\frac{c^2 - v^2}{c^2}\right)^{-\frac{1}{2}}$$

$$\frac{m}{m_0} = \left(\frac{c^2}{c^2 - v^2}\right)^{\frac{1}{2}}$$

$$\frac{m^2}{m_0^2} = \frac{c^2}{c^2 - v^2}$$

$$m^2 c^2 - m^2 v^2 = m_0^2 c^2$$

$$-m^2 v^2 = m_0^2 c^2 - m^2 c^2$$

$$v^2 = \frac{c^2(m_0^2 - m^2)}{-m^2}$$

$$v^2 = \frac{c^2(m^2 - m_0^2)}{m^2}$$

$$v = c\left(1 - \frac{m_0^2}{m^2}\right)^{\frac{1}{2}}$$

Chapter 3 The Idea of a Function

33. $d = \sqrt{\left(0-(-1)\right)^2 + (4-3)^2} = \sqrt{2}$

$p = \left(\frac{0-1}{2}, \frac{4+3}{2}\right) = \left(-\frac{1}{2}, \frac{7}{2}\right)$

35. $d = \sqrt{\left(\frac{1}{2}-\frac{1}{2}\right)^2 + \left(\frac{1}{2}-(-\frac{3}{4})\right)^2} = \frac{5}{4}$

$p = \left(\frac{\frac{1}{2}+\frac{1}{2}}{2}, \frac{\frac{1}{2}-\frac{3}{4}}{2}\right) = \left(\frac{1}{2}, -\frac{1}{8}\right)$

37. $d = \sqrt{(.5-6.2)^2 + (1.6-7.5)^2} = 8.2$

$p = \left(\frac{.5+6.2}{2}, \frac{1.6+7.5}{2}\right) = (3.35,\ 4.55)$

39. $d = \sqrt{\left(2-(-\sqrt{3})\right)^2 + \left(-6-(-3)\right)^2} = \sqrt{4 + 4\sqrt{3} + 3 + 9} = \sqrt{16 + 4\sqrt{3}} = 2\sqrt{4 + \sqrt{3}}$

$p = \left(\frac{2-\sqrt{3}}{2}, \frac{-6-3}{2}\right) = \left(\frac{2-\sqrt{3}}{2}, -\frac{9}{2}\right)$

41. $d = \sqrt{\left(\sqrt{x}-(-\sqrt{x})\right)^2 + \left(\sqrt{y}-(-\sqrt{y})\right)^2} = \sqrt{(2\sqrt{x})^2 + (2\sqrt{y})^2} = \sqrt{4x + 4y} = 2\sqrt{x + y}$

43. $d = \sqrt{(x-x)^2 + \left(y-(-y)\right)^2} = \sqrt{(2y)^2} = 2|y|$

45. $\sqrt{(-4-0)^2 + (-1-y)^2} = \sqrt{(-1-0)^2 + (2-y)^2}$

$16 + 1 + 2y + y^2 = 1 + 4 - 4y + y^2$

$6y = -12$

$y = -2$

11. map distance $= x$

real distance $= y$

$$y = 20x$$

$$y = 20(12.6) = 252$$

13. sodium $= x$

chlorine $= 100 - x$

$$\frac{x}{100 - x} = \frac{5}{3}$$

$$3x = 500 - 5x$$

$$8x = 500$$

$$x = 62.5$$

$$100 - x = 37.5$$

15. $x, \ 8 - x$

$$\frac{x}{8 - x} = \frac{5}{4}$$

$$4x = 40 - 5x$$

$$9x = 40$$

$$x = 4.44$$

$$8 - x = 3.56$$

17. dealer's cost $= x$

list cost $= l$

$$l = 1.18x$$

$$8266 = 1.18x$$

$$x = \$7005$$

21. $y = kx$

$$5 = k10$$

$$k = \frac{1}{2}$$

$$y = \frac{1}{2}x$$

$$y = \frac{1}{2}(12) = 6$$

23. $z = \frac{k}{t}$

$$4 = \frac{k}{\frac{1}{2}}$$

$$k = 2$$

$$z = \frac{2}{t}$$

$$z = \frac{2}{2} = 1$$

25. $v = kt$

$$96 = k3$$

$$k = 32$$

$$v = 32t$$

$$v = 32(2) = 64$$

27. cost of production $= x$

circulation $= y$

$$x = \frac{k}{\sqrt{y}}$$

$$.40 = \frac{k}{\sqrt{200,000}}, \qquad k = 178.89$$

$$x = \frac{178.89}{\sqrt{y}}$$

$$x = \frac{178.89}{\sqrt{400,000}} = .283$$

29. $I = \dfrac{k}{s^2}$

$4 = \dfrac{k}{(5)^2}$

$k = 100$

$I = \dfrac{100}{s^2}$

$I = \dfrac{100}{100} = 1$

31. $p = kri^2$

$50 = k50(.5)^2$

$k = \dfrac{1}{(.5)^2} = 4$

$p = 4ri^2$

$p = 4(50)1^2 = 200$

33. $I = \dfrac{kV}{R}$

$.5 = \dfrac{k(10)}{20}$

$k = 1$

$I = \dfrac{V}{R}$

$I = \dfrac{12}{20} = .6$

35. $C = 2000 + kn$

$2200 = 2000 + k(1000)$

$k = .2$

$C = 2000 + .2n$

$C = 2000 + .2(1500) = 2300$

37. $y = kt$, $x = ct$

    (a) $xy = (ct)(kt) = ckt^2$

           xy varies directly as t^2

    (b) $\dfrac{x}{y} = \dfrac{ct}{kt} = \dfrac{c}{k}$

           $\dfrac{x}{y}$ is a constant

    (c) $x + y = ct + kt = (c + k)t$

           $x + y$ varies directly as t

Section 3.3   Functions

21. $f(x) = \sqrt{-x}$

    $-x \geq 0$, $x \leq 0$: Domain

           $y \geq 0$: Range

23. $f(x) = \sqrt{x-1}$

    $x - 1 \geq 0$, $x \geq 1$: Domain

           $y \geq 0$: Range

25. (a) $f(3)=3(3)+1=10$

(b) $f(\pi)=3\pi+1$

(c) $f(z)=3z+1$

(d) $f(x-h)=3(x-h)+1$

(e) $10=3x+1$

$9=3x$

$3=x$

(f) All reals

(g) $\dfrac{f(a+h)-f(a)}{h}=\dfrac{3(a+h)+1-(3a+1)}{h}=\dfrac{3a+3h+1-3a-1}{h}=\dfrac{3h}{h}=3$

27. $f(x)=x^2$

$$\frac{f(a+h)-f(a)}{h}=\frac{(a+h)^2-a^2}{h}=\frac{a^2+2ah+h^2-a^2}{h}=\frac{h(2a+h)}{h}=2a+h$$

29. $f(x)=x^2-2x$

$$\frac{f(a+h)-f(a)}{h}=\frac{(a+h)^2-2(a+h)-(a^2-2a)}{h}$$

$$=\frac{a^2+2ah+h^2-2a-2h-a^2+2a}{h}=\frac{h(2a+h-2)}{h}=2a+h-2$$

31. $f(x)=x^3-1$

$$\frac{f(a+h)-f(a)}{h}=\frac{(a+h)^3-1-(a^3-1)}{h}=\frac{a^3+3a^2h+3ah^2+h^3-1-a^3+1}{h}$$

$$=\frac{h(3a^2+3ah+h^2)}{h}=3a^2+3ah+h^2$$

33. $f(-x)=-3(-x)^5=3\times 5=-f(x)$, odd

35. $f(-x)=2(-x)-(-x)^3=-2x+x^3=-(2x-x^3)=-f(x)$, odd

37. $f(-x)=(-x)^4-3(-x)=x^4-3x$, neither

39. $f(-x)=\sqrt{(-x)^2+4}=\sqrt{x^2+4}=f(x)$, even

41. $g(x) = 2x - 7$

$\quad g(x-4) = 2(x-4) - 7 = 2x - 15$

$\quad g(x-1) = 2(x-1) - 7 = 2x - 9$

$\quad g(x-4) \cdot g(x-1) = (2x-15)(2x-9) = 4x^2 - 48x + 135$

43. $F(x) = 3x^2$

$\quad F(x-3) = 3(x-3)^2 = 3(x^2 - 6x + 9) = 3x^2 - 18x + 27$

$\quad F(x) \cdot F(x-3) = 3x^2(3x^2 - 18x + 27) = 9x^4 - 54x^3 + 81x^2$

45. $f(x) = x^2 + 2x$

$\quad f(x) - 3 = x^2 + 2x - 3 = 0$

$\qquad (x+3)(x-1) = 0$

$\quad x = 1, \ -3$

47. $f(x) = 2x - x^2$

$\quad f(x) - f(2x) = 2x - x^2 - \left((2)(2x) - (2x)^2 \right) = 0$

$\qquad\qquad\qquad 2x - x^2 - 4x + 4x^2 = 0$

$\qquad\qquad\qquad\qquad\qquad 3x^2 - 2x = 0$

$\qquad\qquad\qquad\qquad\qquad x(3x - 2) = 0$

$\qquad\qquad\qquad\qquad\qquad\qquad x = 0, \ \tfrac{2}{3}$

49. $f(x) = \frac{k}{x}$

(a) $\dfrac{f(x_1)}{f(x_2)} = \dfrac{\frac{k}{x_1}}{\frac{k}{x_2}} = \dfrac{k}{x_1} \cdot \dfrac{x_2}{k} = \dfrac{x_2}{x_1}$

(b) $f(\tfrac{1}{x}) = \dfrac{k}{\frac{1}{x}} = kx$

$\quad \dfrac{1}{f(x)} = \dfrac{1}{\frac{k}{x}} = \dfrac{x}{k}$

(c) $f(x^2) = \dfrac{k}{x^2}$

$\quad \left[f(x) \right]^2 = (\tfrac{k}{x})^2 = \dfrac{k^2}{x^2}$

(d) $\dfrac{f(a+h) - f(a)}{h} = \dfrac{\frac{k}{a+h} - \frac{k}{a}}{h} = \dfrac{\frac{ka - k(a+h)}{(a+h)a}}{h} = \dfrac{-kh}{(a+h)ah} = \dfrac{-k}{(a+h)a}$

(problem #49, continued)

(e) $f(x)+1=\frac{k}{x}+1=\frac{k+x}{x}$

$f(x+1)=\frac{k}{x+1}$

(f) $f(x_1+x_2)=\frac{k}{x_1+x_2}$

$f(x_1)+f(x_2)=\frac{k}{x_1}+\frac{k}{x_2}=\frac{kx_2+kx_1}{x_1x_2}$

(g) $af(x)=a(\frac{k}{x})=\frac{ak}{x}$

$f(ax)=\frac{k}{ax}$

51. $f(s)=3.101s^2+29.46s$

$f(-.415)=3.101(-.415)^2+29.46(-.415)$

$=-11.6918$

53. $i(t)=1.196t^2+.076t$

$i(1.375)=1.196(1.375)^2+.076(1.375)$

$=2.366$

55. $p=w-.12w-.10w-.06w-.015w-150$

$=.705w-150$

Section 3.4   The Graph of a Function

17. $p=z^2-z-6$

$0=(z-3)(z+2)$

$z=3,\ -2,\ (\text{zeros})$

19. $y=\sqrt{16-4x^2}$

$16-4x^2\ge 0$

$(4-2x)(4+2x)\ge 0$

$4-2x$ $\quad + \quad\quad + \quad\quad -$

$4+2x$ $\quad - \quad\quad + \quad\quad +$

$-2 \qquad 2$

Domain: $-2\le x\le 2$

Range: $4\ge y\ge 0$

23. $y(x)=\sqrt{x-2}$

$-2\ge 0,\ x\ge 2:$ Domain
$y\ge 0:$ Range

$0=\sqrt{x-2}$

$x=2,\ (\text{zero})$

25. $v = s^3 - 4s^2$

$0 = s^2(s-4)$

$s = 0, 4$ (zeros)

Section 3.5   Composite and Inverse Functions

1. $f \circ g = f(3-2)x = (3-2x)-5 = -2-2x$

$g \circ f = g(x-5) = 3-2(x-5) = 13-2x$

3. $f \circ g = f(2x) = 3(2x)+4 = 6x+4$

$g \circ f = g(3x+4) = 2(3x+4) = 6x+8$

5. $f \circ g = f(x^2) = x^2$

$g \circ f = g(x) = x^2$

7. $f \circ g = f\left(x^{1/3}\right) = \left(x^{1/3}\right)^3 - 1 = x-1$

$g \circ f = g(x^3-1) = (x^3-1)^{1/3}$

9. $2x+3 = 6,$

$2x = 3$

$x = \frac{3}{2}$

11. $10+4t = 20$

$4t = 10$

$t = \frac{5}{2}$

13.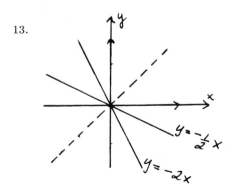

15.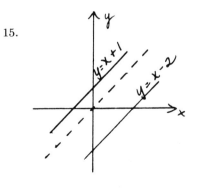

17.

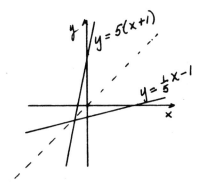

19.

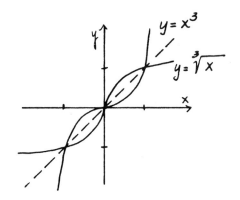

21. $f \circ g = f(\frac{-x}{2}) = -2(\frac{-x}{2}) = x$

23. $f \circ g = f(\frac{-x}{3} + 2) = 6 - 3(\frac{-x}{3} + 2) = x$

25. $f \circ g = f(\frac{x}{5} - 1) = 5\left((\frac{x}{5} - 1) + 1\right) = x$

27. $f\left(g(x)\right) = x$

$g(x) - 3 = x$

$g(x) = x + 3$

29. $f\left(g(x)\right) = x$

$\frac{1}{g(x) + 1} = x$

$g(x) + 1 = \frac{1}{x}$

$g(x) = \frac{1}{x} - 1$

$= \frac{1 - x}{x}$

31. $f\left(g(x)\right) = x$

$\frac{g(x) - 1}{g(x) + 1} = x$

$g(x) - 1 = xg(x) + x \; +1$

$g(x) - xg(x) = x + 1$

$g(x) = \frac{x + 1}{1 - x}$

3. $d=\sqrt{\left(3-(-2)\right)^2+\left(7-(-1)\right)^2}=\sqrt{5^2+8^2}=\sqrt{89}$

5. $p=\frac{F}{a}$, $p=\frac{16}{2.4}=6.67$

7. $C=\frac{v}{n}$, $C=\frac{300,000}{750,000}=.40$

9. $B=\frac{kwd^2}{h}$, $2500=\frac{k(6)(8^2)}{12}$, $k=\frac{2500}{32}=\frac{625}{8}$

   $B=\dfrac{\frac{625}{8}(6)(8^2)}{15}=2000$

23. $g(x)=\sqrt{x-3}$, $x-3\geq0$, $\begin{matrix}x\geq3:&\text{Domain}\\g(x)\geq0:&\text{Range}\end{matrix}$

25. $y=4-2x$, $4-2x=0$, $x=2$

27. $F(x)=-\sqrt{x}+2$, $0=-\sqrt{x}+2$, $\sqrt{x}=2$, $x=4$

29. $y=\sqrt{-x}$, $0=\sqrt{-x}$, $x=0$

33. $f\left(g(x)\right)=x$

   $2g(x)+5=x$

   $g(x)=\frac{x-5}{2}$

35. $g\left(h(x)\right)=x$

   $\dfrac{1}{h(x)+2}=x$

   $1=xh(x)+2x$

   $h(x)=\frac{1}{x}-2=\frac{1-2x}{x}$

37. $f\circ g=f(4-x)=3(4-x)-1=11-3x$

   $g\circ f=g(3x-1)=4-(3x-1)=5-3x$

39. $f\circ g=f(\frac{1}{x})=\dfrac{1}{\frac{1}{x}-1}=\dfrac{1}{\frac{1-x}{x}}=\dfrac{x}{1-x}$

   $g\circ f=g\left(\frac{1}{x-1}\right)=\dfrac{1}{\frac{1}{x-1}}=x-1$

Chapter 4　Elementary Functions

Section 4.1　Linear Functions

13. $m = \frac{2-4}{1-5} = \frac{-2}{-4} = \frac{1}{2}$

15. $m = \frac{-1-(-6)}{-1-3} = \frac{5}{-4} = -\frac{5}{4}$

17. $m = \frac{-3-(-7)}{-2-(-5)} = \frac{4}{3}$

19. $m = \frac{3-3}{-2-5} = 0$

21. $y-5 = \frac{1}{2}(x-2)$

$2y-10 = x-2$

$-8 = x-2y$

23. $y-(-2) = -7(x-5)$

$y+2 = -7x+35$

$y+7x = 33$

25. $y-3 = \left(\frac{3-2}{1-6}\right)(x-1)$

$y-3 = -\frac{1}{5}(x-1)$

$5y-15 = -x+1$

$5y+x = 16$

27. $y-2 = \left(\frac{-1-2}{-1-1}\right)(x-1)$

$y-2 = \frac{3}{2}(x-1)$

$2y-4 = 3x-3$

$2y-3x = 1$

29. $y-2 = \left(\frac{2-0}{0-(-5)}\right)(x-0)$

$y-2 = \frac{2}{5}x$

$5y-10 = 2x$

$5y-2x = 10$

31. $2x-3y = 5$

$3y = 2x-5$

$y = \frac{2x}{3} - \frac{5}{3}$

$m = \frac{2}{3},\ b = -\frac{5}{3}$

33. $x+y = 2$

$y = -x+2$

$m = -1,\ b = 2$

35. $4y-2x+8 = 0$

$4y = 2x-8$

$y = \frac{1}{2}x-2$

$m = \frac{1}{2},\ b = -2$

37. $2y = x+5$

$\quad\quad y = \frac{x}{2} + \frac{5}{2}$

$\quad\quad m = \frac{1}{2},\ b = \frac{5}{2}$

41. $x - 5y + 7 = 0$

$\quad\quad 5y = x + 7$

$\quad\quad y = \frac{x}{5} + \frac{7}{5}$

$\quad\quad m = \frac{1}{5},\ b = \frac{7}{5}$

43. $3x + 2y = 7$

$\quad\quad 2y = -3x + 7$

$\quad\quad y = -\frac{3}{2}x + \frac{7}{2}$

$\quad\quad m = -\frac{3}{2},\ m' = \frac{2}{3}$

$\quad\quad y - 2 = \frac{2}{3}(x-1)$

$\quad\quad 3y - 6 = 2x - 2$

$\quad\quad 3y - 2x = 4$

45. $x - y = 2$

$\quad\quad y = x - 2$

$\quad\quad m = 1,\ m' = -1$

$\quad\quad y - 3 = -1(x-5)$

$\quad\quad y + x = 8$

47. $2y - 3x = 1$

$\quad\quad 2y = 3x + 1$

$\quad\quad y = \frac{3}{2}x + \frac{1}{2}$

$\quad\quad m = \frac{3}{2},\ m' = -\frac{2}{3}$

$\quad\quad y - 2 = -\frac{2}{3}(x-1)$

$\quad\quad 3y - 6 = -2x + 2$

$\quad\quad 3y + 2x = 8$

49. $5x - 7y = 0$

$\quad\quad 7y = 5x$

$\quad\quad y = \frac{5}{7}x$

$\quad\quad m = \frac{5}{7},\ m' = -\frac{7}{5}$

$\quad\quad y - 0 = -\frac{7}{5}(x-0)$

$\quad\quad 5y = -7x$

$\quad\quad 5y + 7x = 0$

51. $-x + 3y = 5$

$\quad\quad 3y = x + 5$

$\quad\quad y = \frac{x}{3} + \frac{5}{3}$

$\quad\quad m = \frac{1}{3},\ m' = -3$

$\quad\quad y - 1 = -3(x+2)$

$\quad\quad y - 1 = -3x - 6$

$\quad\quad y + 3x = -5$

53. $Ax + By = C$

$\quad\quad By = -Ax + C$

$\quad\quad y = \frac{-A}{B}x + \frac{C}{B}$

$\quad\quad m = \frac{-A}{B},\ m' = \frac{B}{A}$

$\quad\quad y - y_1 = \frac{B}{A}(x - x_1)$

$\quad\quad Ay - Ay_1 = Bx - Bx_1$

$\quad\quad Bx_1 - Ay_1 = Bx - Ay$

55. $x+3y=2$

If $x=0$, $y=\frac{2}{3}$

If $y=0$, $x=2$

$$\frac{x}{2}+\frac{y}{\frac{2}{3}}=1$$

Section 4.2 Mathematical Models

1a. $C=5(300)+400=1900$

1b. $A=\frac{C}{x}=\frac{1900}{300}=6.33$

1c. $A=\frac{5(500)+400}{500}=\frac{2900}{500}=5.80$

1d. $m=5$

1e. $2000=5x+400$

$5x=1600$

$x=\frac{1600}{5}=320$

3a. $m=2$

3b. $v=2(100)=200$

3c. $b=300$, unchanged

3d. $100=2x+300$

$2x=-200$

$x=-100$, No items

$500=2x+300$

$2x=200$

$x=100$

$1000=2x+300$

$2x=700$

$x=350$

5a. $R=ux=30(2810)=84.300$

5b. $P=ux-mx-b=30(2810)-8(2810)-24{,}000=37{,}820$

5c. $P=(30-8)x-24{,}000$

$O=22x-24000$

$X=\frac{24000}{22}=1090.91\simeq1091$

7a. $v=.37x=.37(50{,}000)=18{,}500$

7b. $P=(u-m)x-b$

$=(.62-.37)(50{,}000)-18{,}000=.25(50{,}000)-18{,}000=-5500$ (a loss)

7c. $O=.25x-18{,}000$

$X=\frac{18000}{.25}=72{,}000$

9. $I_1 = \frac{1}{2}$, $v_1 = 6$

$I_2 = \frac{2}{3}$, $v_2 = 8$

$I - \frac{1}{2} = \left(\dfrac{\frac{1}{2} - \frac{2}{3}}{6 - 8} \right)(v - 6)$

$I - \frac{1}{2} = \frac{1}{12}(v - 6)$

$12I - 6 = v - 6$

$I = \frac{v}{12}$

11. $V = C - \dfrac{Cn}{N}$

$V = 300 - \dfrac{300(5)}{20} = 225$

13. $I = Art = 4000(.12)3 = 1440$

$P = \dfrac{4000 + 1440}{35} = 151.11$

Section 4.3   Quadratic Functions: The Parabolic Graph

1. $f(0) = 0 - 4 = -4$

$x^2 - 4 = 0$

$(x - 2)(x + 2) = 0$, $x = \pm 2$

vertex: $(0, -4)$

3. $f(0) = 1$

$x^2 + 1 = 0$, no x-intercepts

vertex: $(0, 1)$

5. $f(0) = 9$

$9 - x^2 = 0$

$(3 - x)(3 + x) = 0$, $x = \pm 3$

vertex: $(0, 9)$

7. $x(0) = 0$

$t^2 - 3t = 0$

$t(t - 3) = 0$, $t = 0, 3$

vertex: $\left(\frac{3}{2}, f\left(\frac{3}{2}\right) \right) = \left(\frac{3}{2}, \frac{-9}{4} \right)$

9. $f(0) = 6$

$x^2 - 5x + 6 = 0$

$(x - 3)(x - 2) = 0$, $x = 2, 3$

vertex: $\left(\frac{5}{2}, f\left(\frac{5}{2}\right) \right) = \left(\frac{5}{2}, \frac{-1}{4} \right)$

11. $f(0) = 1$

$3x^2 + 2x + 1 = 0$

$x = \dfrac{-2 \pm \sqrt{4 - 12}}{6}$

no real roots

vertex: $\left(\frac{-2}{6}, f\left(\frac{-2}{6}\right) \right) = \left(\frac{-1}{3}, \frac{2}{3} \right)$

13. $f(0) = 2$

$-x^2 + x + 2 = 0$

$x^2 - x - 2 = 0$

$(x - 2)(x + 1) = 0$, $x = -1, 2$

vertex: $\left(\frac{1}{2}, f\left(\frac{1}{2}\right) \right) = \left(\frac{1}{2}, \frac{9}{4} \right)$

15. $f(0) = -5$

$2x^2 - 3x - 5 = 0$

$(2x - 5)(x + 1) = 0$, $x = -1, \frac{5}{2}$

vertex: $\left(\frac{3}{4}, f\left(\frac{3}{4}\right) \right) = \left(\frac{3}{4}, \frac{-49}{8} \right)$

17. $f(0)=3$

    $3+x-x^2 = 0$

    $x^2-x-3 = 0$

$$x = \frac{1\pm\sqrt{1-4(1)(-3)}}{2} = \frac{1\pm\sqrt{13}}{2}$$

    vertex: $\left(\frac{1}{2},\ f\left(\frac{1}{2}\right)\right)=\left(\frac{1}{2},\ \frac{13}{4}\right)$

19. $G(0)=1$

    $x^2+x+1 = 0$

$$x = \frac{-1\pm\sqrt{1-4}}{2},\ \text{no real roots}$$

    vertex: $\left(\frac{-1}{2},\ f\left(\frac{-1}{2}\right)\right)=\left(\frac{-1}{2},\ \frac{3}{4}\right)$

21. $y = k(x-0)(x-2)$

    $-1 = k(1)(1-2)$

    $k = 1,$

    $y = x(x-2)= x^2-2x$

23. $y = k(x-1)^2$

    $1 = k(-1)^2$

    $k = 1$

    $y = (x-1)^2$

25. $y = k(x+1)(x+3)$

    $1 = k(-2+1)(-2+3)$

    $k = -1$

    $y = -(x+1)(x+3)= -x^2-4x-3$

27. $y = k(x+1)(x-3)$

    $-2 = k(1)(-3)$

    $k = \frac{2}{3}$

    $y = \frac{2}{3}(x+1)(x-3)$

    $y = \frac{2x^2}{3}-\frac{4x}{3}-2$

29. $y = k(x-2)(x+5)$

    $1 = k(-2)(5)$

    $k = -\frac{1}{10}$

    $y = -\frac{1}{10}(x-2)(x+5)$

    $= -\frac{1}{10}x^2-\frac{3x}{10}+1$

35b. $P(t)=45t-3t^2$

    $45t-3t^2 > 0$

    $3t(15-t) > 0$

$3t$	$-$	$+$	$+$
$15-t$	$+$	$+$	$-$

    0   15

  c. $P(t)<0$ for $15<t<24$

37. $v= 100+200t-25t^2$

    vertex: $\left(\frac{-200}{-50},\ v(4)\right)= (4,500)$

    $v_{max} = 500$

    $100+200t-25t^2 = 0$

$$t = \frac{-200\pm\sqrt{200^2-4(-25)(100)}}{2(-25)} = 4\pm4.47$$

    $t = 8.47$

39. $f \circ g = f(ax^2 + bx + c) = A(ax^2 + bx + c) + B(ax^2 + bx + c) + C$

$= A\left(a^2x^4 + 2abx^3 + (2ac + b)x^2 + 2bcx + c^2\right) + B(ax^2 + bx + c) + C$ is not a quadratic

Section 4.4   Polynomial Functions

1. $y = x^3 + 1 = (x+1)(x^2 - x + 1) = 0,\ x = -1$

$f(0) = 1$

$$\begin{array}{ccc} -+ & ++ \\ \hline & \rlap{\;+} \\ & -1 \end{array}$$

3. $y = 1 - x^4 = (1 + x^2)(1 - x)(1 + x) = 0,\ x = \pm 1$

$f(0) = 1$

$$\begin{array}{ccc} ++- & +++ & +-+ \\ \hline & -1 & \quad 1 \end{array}$$

5. $y - 3 = (x-1)^3$

Translate $(0, 0)$ of $y = x^3$ to $(1, 3)$

7. $y + 2 = (x+1)^5$

Translate $(0, 0)$ of $y = x^5$ to $(-1, -2)$

9. $y - 1 = -(x-2)^4$

Translate $(0, 0)$ to $y = -x^4$ to $(2, 1)$

11. $y = x(x-1)(x+1) = 0,\ x = 0,\ \pm 1$

$f(0) = 0$

$$\begin{array}{cccc} --- & --+ & +-+ & +++ \\ \hline & -1 & 0 & 1 \end{array}$$

13. $y = x^2(x-1)(x+1) = 0,\ x = 0,\ \pm 1$

$f(0) = 0$

$$\begin{array}{cccc} +-- & +-+ & +-+ & +++ \\ \hline & -1 & 0 & 1 \end{array}$$

15. $y = x^3(x-2)(x+2) = 0,\ x = 0,\ \pm 2$

$f(0) = 0$

$$\begin{array}{cccc} --- & --+ & +-+ & +++ \\ \hline & -2 & 0 & 2 \end{array}$$

17. $y = 3(x^3 - 27) = 3(x-3)(x^2 + 3x + 9) = 0$, $x = 3$

$f(0) = -81$

```
      + - +   + + +
    ─────────┼─────────
             3
```

19. $y = (x^2 - 1)^2 = (x-1)^2(x+1)^2 = 0$, $x = \pm 1$

$f(0) = 1$

```
    + +   + +   + +
   ──────┼─────┼──────
        -1     1
```

21. $y = x(x-1)(x+3) = 0$, $x = 0, 1, -3$

$f(0) = 0$

```
    - - -   - - +   + - +   + + +
   ────────┼───────┼───────┼───────
          -3       0       1
```

23. $y = (x+1)^2 x(x-1) = 0$, $x = -1, 0, 1$

$f(0) = 0$

```
    + - -   + - -   + + -   + + +
   ────────┼───────┼───────┼───────
          -1       0       1
```

25. $y = (x^2 - 9)(x^2 + 4) = (x-3)(x+3)(x^2 + 4) = 0$, $x = \pm 3$

$f(0) = (-9)(4) = -36$

```
    - - +   - + +   + + +
   ────────┼───────┼───────
          -3       3
```

27. $y = (x+2)(x-3)(x+5) = 0$, $x = -2, 3, -5$

$f(0) = -30$

```
    - - -   - - +   + - +   + + +
   ────────┼───────┼───────┼───────
          -5       -2      3
```

29. $y = (x-5)(x-1)(x+2)^2 = 0$, $x = 5, 1, -2$

$f(0) = (-5)(-1)(4) = 20$

```
    - - +   - - +   - + +   + + +
   ────────┼───────┼───────┼───────
          -2       1       5
```

3. $p = \dfrac{3x - 2}{x - 4}$

   Vert. A.: $x - 4 = 0$, $x = 4$

   Horiz. A.: $y = \dfrac{3 - \frac{2}{x}}{1 - \frac{4}{x}}$, as $x \to \infty$, $y \to \dfrac{3}{1}$, $y = 3$

   Intercepts: $f(0) = \dfrac{-2}{-4} = \dfrac{1}{2}$, $\left(0, \dfrac{1}{2}\right)$

   $\qquad\qquad 3x - 2 = 0$, $x = \dfrac{2}{3}$, $\left(\dfrac{2}{3}, 0\right)$

5. $p = \dfrac{1}{w + 1}$

   Vert. A.: $w + 1 = 0$, $w = -1$

   Horiz. A.: $p = \dfrac{\frac{1}{w}}{1 + \frac{1}{w}}$, as $w \to \infty$, $p = 0$

   Intercepts: $p(0) = 1$, $(0, 1)$

   $\qquad\qquad p = 0 = \dfrac{1}{w + 1}$, no w-intercept

7. $s = \dfrac{5w^2}{4 - w^2}$

   Vert. A.: $4 - w^2 = (2-w)(2+w) = 0$, $w = \pm 2$

   Horiz. A.: $s = \dfrac{5}{\frac{4}{w^2} - 1}$, as $w \to \infty$, $s = -5$

   Intercepts: $s(0) = 0$, $(0, 0)$

   $\qquad\qquad 0 = \dfrac{5w^2}{4 - w^2}$, $(0, 0)$

9. $y = \dfrac{x^2}{x^2 - 9}$

Vert. A.: $\quad x^2 - 9 = (x{-}3)(x{+}3)=0,\ x = \pm 3$

Horiz. A.: $\quad y = \dfrac{1}{1 - \dfrac{9}{x^2}}$, as $x \to \infty,\ y = 1$

Intercepts: $y(0) = 0,\ (0,\ 0)$

$$x^2 = 0,\ x = 0,\ (0,\ 0)$$

11. $s = \dfrac{(t{+}2)^2}{t(t{+}2)} = \dfrac{t^2 + 4t + 4}{t^2 + 2t} = \dfrac{t + 2}{t}$

Vert. A.: $\quad t = 0$

Horiz. A.: $\quad s = \dfrac{1 + \dfrac{4}{t} + \dfrac{4}{t^2}}{1 + \dfrac{2}{t}}$, as $t \to \infty,\ s = 1$

Intercepts: $s(0)$ does not exist

$$s = \dfrac{(t{+}2)^2}{t(t{+}2)} = 0,\ t = -2,\ s(-2) \text{ does not exist}$$

13. $y = \dfrac{3}{w^2 - 9w} = \dfrac{3}{w(w{-}9)}$

Vert. A.: $\quad w(w{-}9) = 0,\ w = 0,\ 9$

Horiz. A.: $\quad y = \dfrac{\dfrac{3}{w^2}}{1 - \dfrac{9}{w}}$, as $w \to \infty,\ y = 0$

Intercepts: $y(0)$ does not exist

$$y = 0 = \dfrac{3}{w(w{-}9)}, \text{ no solution}$$

15. $y = \dfrac{1}{(p{-}4)(p{+}1)}$

Vert. A.: $\quad (p{-}4)(p{+}1) = 0,\ p = -1,\ 4$

Horiz. A.: $\quad y = \dfrac{\dfrac{1}{p^2}}{1 - \dfrac{3}{p} - \dfrac{4}{p^2}}$, as $p \to \infty,\ y = 0$

Intercepts: $y(0) = \dfrac{1}{(-4)(1)} = -\dfrac{1}{4}$

$$y = 0 = \dfrac{1}{(p{-}4)(p{+}1)}, \text{ no solution}$$

17. $y = \dfrac{5x + 3}{x(x+1)}$

   Vert. A.: $x(x+1) = 0$, $x = 0, -1$

   Horiz. A.: $y = \dfrac{\frac{5}{x} + \frac{3}{x^2}}{1 + \frac{1}{x}}$, as $x \to \infty$, $y = 0$

   Intercepts: $y(0) =$ does not exist

   $$y = 0 = \dfrac{5x + 3}{x(x+1)}, \ \ 0 = 5x + 3, \ x = -\dfrac{3}{5}$$

21. $u = \dfrac{v^2 + 3}{4v^2 + 5}$

   Vert. A.: $4v^2 + 5 = 0$, no solution

   Horiz. A.: $u = \dfrac{1 + \frac{3}{v^2}}{4 + \frac{5}{v^2}} = \dfrac{1}{4}$

   Intercepts: $u(0) = \dfrac{3}{5}$

   $$u = 0 = \dfrac{v^2 + 3}{4v^2 + 5}, \text{ no solution}$$

Section 4.6   Multipart Functions

17. $f(x) = |x+1| = \begin{cases} x+1, & x \geq -1 \\ -(x+1), & x < -1 \end{cases}$

19. $H(x) = |2x| = \begin{cases} 2x, & x \geq 0 \\ -2x, & x < 0 \end{cases}$

21. $f(x) = |x| - 3 = \begin{cases} x-3, & x \geq 0 \\ -x-3, & x < 0 \end{cases}$

23. $y = |x| + 3 = \begin{cases} x+3, & x \geq 0 \\ -x+3, & x < 0 \end{cases}$

25. $f(t) = |t^2| = t^2$

27. $f(x) = x^2 - |x| + 1 = \begin{cases} x^2 - x + 1, & x \geq 0 \\ x^2 + x + 1, & x < 0 \end{cases}$

29. $g(x) = |2x^2 + x - 1| = |(2x-1)(x+1)|$

   $(2x-1)(x+1) = 0$, $x = -1, \frac{1}{2}$

$2x-1$	$-$	$-$	$+$
$x+1$	$-$	$+$	$+$

   $$ \overline{\underset{-1}{+}\underset{\frac{1}{2}}{+}}$$

   $$g(x) = \begin{cases} 2x^2 + x - 1 & x \leq -1 \\ -(2x^2 + x - 1) & -1 < x < \frac{1}{2} \\ 2x^2 + x - 1 & x \geq \frac{1}{2} \end{cases}$$

– 57 –

31. $y = x^2 + 3|x| + 2 = \begin{cases} x^2 + 3x + 2 & x \geq 0 \\ x^2 - 3x + 2 & x < 0 \end{cases}$

33. $f(x) = (x-1)|x+1| = \begin{cases} (x-1)(x+1), & x \geq 1 \\ -(x-1)(x+1), & x < -1 \end{cases}$

35. $f(x) = x|x-4| = \begin{cases} x(x-4), & x \geq 4 \\ -x(x-4), & x < 4 \end{cases}$

Chapter 4   Review

1. $m = \dfrac{3-(-2)}{-1-0} = -5$

3. $m = \dfrac{2-(-5)}{3-1} = \dfrac{7}{2}$

7. $m = \dfrac{4-(-1)}{5-(-1)} = \dfrac{5}{6}, \ m' = -\dfrac{6}{5}$

9. $y - (-1) = -2(x-3)$
$$y + 1 = -2x + 6$$
$$y + 2x = 5$$

11. $y - 2 = \left(\dfrac{2-4}{7-(-1)}\right)(x-7)$
$$y - 2 = -\tfrac{1}{4}(x-7)$$
$$4y - 8 = -x + 7$$
$$4y + x = 15$$

13. $2x + 7y = -3$
$$7y = -2x - 3$$
$$y = -\tfrac{2}{7}x - \tfrac{3}{7}$$
$$m = -\tfrac{2}{7}x, \ b = -\tfrac{3}{7}$$

15. $3x - 6y = 1$
$$6y = 3x - 1$$
$$y = \tfrac{1}{2}x - \tfrac{1}{6}$$
$$m = \tfrac{1}{2}, \ b = -\tfrac{1}{6}$$

17. $x - 2y = -5$
$$2y = x + 5$$
$$y = \tfrac{1}{2}x + \tfrac{5}{2}$$
$$m = \tfrac{1}{2}, \ b = \tfrac{5}{2}, \ m' = -2$$
$$y - \tfrac{5}{2} = -2(x-0)$$
$$2y - 5 = -4x$$
$$2y + 4x = 5$$

19. $x - 3y = 7$
$$3y = x - 7$$
$$y = \tfrac{1}{3}x - \tfrac{7}{3}$$
$$m = \tfrac{1}{3}, \ b = -\tfrac{7}{3}, \ m' = -3$$
$$y - (-2) = -3(x-1)$$
$$y + 2 = -3x + 3$$
$$y + 3x = 1$$

21. $y = x^2 + 7$
$$y(0) = 7, \ \text{vertex:} \quad (0, \ 7)$$

23. $f(x)=2x^2-x-3$

    $f(0)=-3$

    $(2x-3)(x+1)=0,\ x=\frac{3}{2},\ -1$

    vertex: $\left(\frac{1}{4},\ f\left(\frac{1}{4}\right)\right)=\left(\frac{1}{4},\ \frac{-25}{8}\right)$

25. $s=2+t-t^2$

    $s(0)=2$

    $2+t-t^2=(2-t)(1+t)=0,\ t=-1,\ 2$

    vertex: $\left(\frac{1}{2},\ f\left(\frac{1}{2}\right)\right)=\left(\frac{1}{2},\ \frac{9}{4}\right)$

27. x-intercepts: $(-2,\ 0),\ (6,\ 0)$

    $y=k(x+2)(x-6)$

    y-intercepts: $(0,\ 1)$

    $1=k(2)(-6)$

    $k=-\frac{1}{12}$

    $y=-\frac{1}{12}(x+2)(x-6)$

29. x-intercepts: $(-1,\ 0)(-3,\ 0)$

    $y=k(x+1)(x+3)$

    y-intercept: $(0,\ 2)$

    $2=k(1)(+3)$

    $k=\frac{2}{3}$

    $y=\frac{2}{3}(x+1)(x+3)$

31. $y=2-x^3=0,\ x^3=2,\ x=\sqrt[3]{2}$

    $f(0)=2$

33. $y=x(x+1)(x+5)=0,\ x=0,\ -1,\ 5$

    $f(0)=0$

35. $y=(x-2)(x+1)^2=0,\ x=2,\ -1$

    $f(0)=(-2)(1)=-2$

37. $y=\frac{3}{x+2}$

    Vert. A: $x+2=0,\ x=-2$

    Horiz. A: $y=\frac{3}{x+2}=0$, no solution

    Intercepts: $y(0)=\frac{3}{2},\ \left(0,\ \frac{3}{2}\right)$

    $0=\frac{3}{x+2}$, no solution

39. $p = \dfrac{2x}{(x-2)(x+2)}$

Vert. A: $(x-2)(x+2) = 0$, $x = \pm 2$

Horiz. A: $p = \dfrac{\frac{2}{x}}{1 - \frac{4}{x^2}}$, as $x \to \infty$, $p = 0$

Intercepts: $p(0) = 0$, $(0, 0)$

$$0 = \dfrac{2x}{(x-2)(x+2)}, \quad x = 0, \ (0, 0)$$

45. $F(x) = \left| x - \dfrac{1}{2} \right| = \begin{cases} x - \dfrac{1}{2}, & x \geq \dfrac{1}{2} \\ -x + \dfrac{1}{2}, & x < \dfrac{1}{2} \end{cases}$

47. $y = |t| + 2 = \begin{cases} t + 2, & t \geq 0 \\ -t + 2, & t < 0 \end{cases}$

49. $s = t^2 + 2t = t(t+2) = 0$, $t = 0, -2$

$s(0) = 0$

vertex: $\left(-\dfrac{2}{2}, f(-1) \right) = (-1, -1)$

51. $y = x^2 - x = x(x-1) = 0$, $x = 0, 1$

$y(0) = 0$

vertex: $\left(\dfrac{1}{2}, f\left(\dfrac{1}{2}\right) \right) = \left(\dfrac{1}{2}, -\dfrac{1}{4} \right)$

57. $v - v_1 = \left(\dfrac{v_1 - v_2}{t_1 - t_2} \right)(t - t_1)$

$v - 25 = \left(\dfrac{25 - 38}{4 - 10} \right)(t - 4)$

$v - 25 = \dfrac{13}{6}(t - 4)$

$v = \dfrac{13t}{6} + \dfrac{98}{6}$

59. $m = \dfrac{-315 - (-225)}{0 - 500} = \dfrac{9}{50}$

61. $P(t) = 10(2t - 3) = 20t - 30$

$P(t) = 20t - 30 > 0$

$20t > 30$, $t > \dfrac{3}{2}$

63a. $P(n) = 12n - 5500 > 0$

63c. $P(n) = 12n - 5500 > 0$

$12n > 5500$, $n > 459$

Chapter 5   Exponential and Logarithmic Functions

Section 5.1   Exponential Functions

1. $x^{-3/2} = \frac{27}{8}$

$\left(x^{-3/2}\right)^{-2/3} = \left(\frac{27}{8}\right)^{-2/3}$

$x = \left(\frac{3}{2}\right)^{-2} = \frac{4}{9}$

3. $27^{x+2} = 9^{2-x}$

$(3^3)^{x+2} = (3^2)^{2-x}$

$3(x+2) = 2(2-x)$

$3x+6 = 4-2x$

$5x = -2$

$x = -\frac{2}{5}$

5. $(-2)^{x+2} = -8$

$(-2)^{x+2} = (-2)^3$

$x+2 = 3$

$x = 1$

7. $\left((3)^x\right)^x = 1$

$3^{x^2} = 1$

$x^2 = 0$

$x = 0$

9. $3^{x^2+4x} = \frac{1}{81}$

$3^{x^2+4x} = (3)^{-4}$

$x^2 + 4x = -4$

$x^2 + 4x + 4 = 0, \ (x+2)^2 = 0, \ x = -2$

41. $S = 500{,}000(.98)^t$

$S(4) = 500{,}000(.98)^4 = 461{,}184$

43. $I = P(1+i)^t - P$

$= 500(1.045)^5 - 500 = 123.09$

45. $P = 1000(1+.035)^{10} = 1410.60$

47. $I = 95{,}000(1+.015)^8 - 5000 = 632.46$

Section 5.2   Exponential Functions with Base e

11. $v = 10.772e^{(2.032)(2.75)} = 10.772e^{5.588} = 10.772(267.2) = 2878.29$

13. $T = 198.6 + 95.3e^{(-.147)(7.5)} = 198.6 + 95.3e^{-1.1025} = 198.6 + 95.3(.3320) = 230.24$

17. $f(t) = \sqrt{t+3.346}\,e^{.0178t}$

   $f(30.542) = \sqrt{30.542+3.346}\,e^{(0.178)(30.542)}$

   $= \sqrt{33.888}\,e^{.54365} = (5.8213)(1.7223) = 10.026$

19. $H(.7683) = \left(e^{(.15)(.7683)} + e^{(-.15)(.7683)}\right)e^{3.19-.7683}$

   $= \left(e^{.11525} + e^{-.11525}\right)e^{2.4217}$

   $= (1.122+.8911)(11.265) = 22.68$

Section 5.3   The Logarithm Function

7. $\log_b 4 = 2$
   $b^2 = 4,\ b = 2$

9. $\log_b 100 = 2$
   $b^2 = 100,\ b = 10$

11. $\log_{10} x = 4$
   $10^4 = x$

13. $\log_x 10 = 1$
   $x^1 = 10$

15. $\log_x 64 = 3$
   $x^3 = 64$
   $x = 4$

17. $\log_{27} x = \frac{2}{3}$
   $27^{2/3} = x$
   $x = 3^2 = 9$

19. $\log_3 9 = x$
   $3^x = 9 = 3^2$
   $x = 2$

21. $\log_b b^a = x$
   $b^x = b^a$
   $x = a$

23. $\log_x 2 = \frac{1}{3}$
   $x^{1/3} = 2$
   $x = 2^3 = 8$

25. $\log_x 6 = \frac{1}{2}$
   $x^{1/2} = 6$
   $x = 6^2 = 36$

29. $\log_2 (x+3) = -1$
   $2^{-1} = x+3$
   $x = \frac{1}{2} - 3 = -\frac{5}{2}$

31. $\log_3 (x+1) < 2$
   $3^2 > x+1$
   $8 > x$
   $x+1>0,\ x>1$
   $1 < x < 8$

33. $2 \leq \log_2 x \leq 3$

$2 \leq \log_2 x \qquad \log_2 x \leq 3$

$2^2 \leq x \qquad\qquad x \leq 2^3$

$\qquad\qquad\qquad 4 \leq x \leq 8$

51. $2 = \log_a 4$

$a^2 = 4, \; a = 2$

53. $-1 = \log_a{}^{(.1)}$

$a^{-1} = \frac{1}{10}$

$a = 10$

57a. $f(x) = \log_2 x$

$\quad f(x+y) = \log_2 (8+8) = \log_2 16 = 4$

$\quad f(x) + f(y) = \log_2 8 + \log_2 8 = 3+3$

$\quad f(x) + f(y) \neq f(x+y)$

57b. $f(ax) = f(4 \cdot 2) = \log_2 (4 \cdot 2) = 3$

$\quad af(x) = 4f(2) = 4 \log_2 2 = 4$

$\quad f(ax) \neq af(x)$

69. $A(t) = A_0 e^{-t/5}$

$\quad \frac{A_0}{2} = A_0 e^{-t/5}$

$\quad \frac{1}{2} = e^{-t/5}$

$\quad \ln \frac{1}{2} = \ln e^{-t/5}$

$\quad \ln \frac{1}{2} = \frac{-t}{5}$

$\quad t = -5 \ln \frac{1}{2} = 3.4657$

Section 5.4 Basic Properties of Logarithms

1. $\log_2 32 \cdot 16 = \log_2 32 + \log_2 16 = \log_2 2^5 + \log_2 2^4 = 5 + 4 = 9$

3. $\log_5 25^{1/4} = \frac{1}{4} \log 5^2 = \frac{1}{4} \cdot 2 = \frac{1}{2}$

5. $\log_3 27 \cdot 9 \cdot 3 = \log_3 3^3 + \log_3 3^2 + \log_3 3 = 3 + 2 + 1 = 6$

7. $\log_2 (8 \cdot 32)^3 = 3\left[\log_2 8 + \log_2 32\right] = 3 \log_2 2^3 + 3 \log_2 2^5 = 9 + 15 = 24$

9. $\log \frac{3}{2} = \log 3 - \log 2 = .1761$

11. $\log 12 = \log 3 \cdot 4 = \log 3 + \log 4 = \log 3 + \log 2^2 = 1.0791$

13. $\log 90 = \log 9 \cdot 10 = \log 9 + \log 10 = \log 3^2 + \log 10 = 1.9542$

15. $\log \sqrt{5} = \log \left(\frac{10}{2}\right)^{1/2} = \frac{1}{2}\left[\log 10 - \log 2\right] = .3495$

17. $\log 2400 = \log 8 \cdot 3 \cdot 10^2 = \log 8 + \log 3 + \log 10^2 = 3.3801$

19. $\log .0014 = \log 2 \cdot 7 \cdot 10^{-4} = \log 2 + \log 7 + \log 10^{-4} = -2.8539$

21. $\log_2 x^2 - \log_2 x = \log_2 \frac{x^2}{x} = \log_2 x$

    or

    $\log_2 x^2 - \log_2 x = 2 \log_2 x - \log_2 x = \log_2 x$

23. $\log x + \log \frac{1}{x} = \log x \cdot \frac{1}{x} = \log 1 = 0$

25. $\log 5t + 2 \log (t^2 - 4) - \frac{1}{2} \log (t+3)$

    $= \log \dfrac{5t(t^2-4)^2}{(t+3)^{1/2}}$

27. $3 \log u - 2 \log (u+1) - 5 \log (u-1)$

    $= \log \dfrac{u^3}{(u+1)^2(u-1)^5}$

29. $\ln I = \left(\dfrac{-R}{L}\right)t + \ln I_0$

    $\ln I - \ln I_0 = \dfrac{-Rt}{L}$

    $\ln \dfrac{I}{I_0} = \dfrac{-Rt}{L}, \dfrac{I}{I_0} = e^{-\frac{R}{L}t}$

Section 5.5   Exponential and Logarithmic Equations

1. $\qquad 7^{x+1} = 2^x$

    $\log 7^{x+1} = \log 2^x$

    $(x+1) \log 7 = x \log 2$

    $x \log 7 + \log 7 = x \log 2$

    $x(\log 7 - \log 2) = -\log 7$

    $x = \dfrac{-\log 7}{\log 7 - \log 2} = -1.553$

3. $\qquad 10^{x^2} = 2^x$

    $\log 10^{x^2} = \log 2^x$

    $x^2 \log 10 = x \log 2$

    $x^2 - x \log 2 = 0$

    $x(x - \log 2) = 0$

    $x = 0, \log 2 = .301$

5.
$$2^{x+1} = 3$$
$$\log 2^{x+1} = \log 3$$
$$(x+1)\log 2 = \log 3$$
$$x \log 2 + \log 2 = \log 3$$
$$x \log 2 = \log 3 - \log 2$$
$$x = \frac{\log 3 - \log 2}{\log 2} = .5850$$

7.
$$2^{x+5} = 3^{x-2}$$
$$\log 2^{x+5} = \log 3^{x-2}$$
$$(x+5)\log 2 = (x-2)\log 3$$
$$x(\log 2 - \log 3) = -2\log 3 - 5\log 2$$
$$x = \frac{-2\log 3 - 5\log 2}{\log 2 - \log 3}$$
$$= 13.97$$

9.
$$2^{x+1} > 5$$
$$\ln 2^{x+1} > \ln 5$$
$$(x+1)\ln 2 > \ln 5$$
$$x \ln 2 > \ln 5 - \ln 2$$
$$x > \frac{\ln 5 - \ln 2}{\ln 2} = 1.32$$

11.
$$2^{\log x} = 2$$
$$\log x = 1$$
$$x = 10$$

13.
$$(\log x)^{1/2} = \log \sqrt{x} = \log x^{1/2}$$
$$(\log x)^{1/2} = \tfrac{1}{2} \log x$$
$$\log x = \tfrac{1}{4} \log^2 x$$
$$\tfrac{1}{4} \log^2 x - \log x = 0$$
$$\log x \left(\tfrac{1}{4} \log x - 1 \right) = 0$$
$$\log x = 0 \qquad \tfrac{1}{4} \log x = 1$$
$$x = 1 \qquad \log x = 4$$
$$x = 10^4$$

15.
$$(\log x)^3 = \log x^3 = 3 \log x$$
$$(\log x)^3 - 3 \log x = 0$$
$$\log x(\log^2 x - 3) = 0$$
$$\log x = 0, \qquad \log x = \sqrt{3}, \qquad \log x = -\sqrt{3}$$
$$x = 1 \qquad x = 10^{\sqrt{3}} \qquad x = 10^{-\sqrt{3}}$$

17. $\log (x+15) + \log x = 2$

$$\log x (x+15) = 2$$

$$x (x+15) = 10^2$$

$$x^2 + 15x - 100 = 0$$

$$(x+20)(x-5) = 0$$

$x = 5$, $x = -20$, impossible, $\log (-20)$ is not defined

19. $\log (x-2) - \log (2x+1) = \log \frac{1}{x}$

$$\log \frac{(x-2)}{(2x+1)} = \log \frac{1}{x}$$

$$\frac{x-2}{2x+1} = \frac{1}{x}$$

$$x(x-2) = 2x+1$$

$$x^2 - 4x - 1 = 0$$

$x = 2+\sqrt{5}$, $x = 2-\sqrt{5}$, impossible, $\log (2-\sqrt{5})$ is not defined

21. $\log (x+2) - \log (x-2) - \log x + \log (x-3) = 0$

$$\log \frac{(x+2)(x-3)}{x(x-2)} = 0$$

$$\frac{(x+2)(x-3)}{x(x-2)} = 10^0 = 1$$

$$x^2 - 2x = x^2 - x - 6$$

$$x = 6$$

23. $\log x + \log (x^2-4) - \log 2x = 0$

$$\log \frac{x(x^2-4)}{2x} = 0$$

$$\frac{x^2-4}{2} = 10$$

$$x^2 - 6 = 0$$

$x = \sqrt{6}$, $x = -\sqrt{6}$, impossible, $\log (-\sqrt{6})$ is not defined

25. $\log (x+1) - \log x < 1$

$$\log \left(\frac{x+1}{x}\right) < 1$$

$$\frac{x+1}{x} < 10^1$$

$$x+1 < 10x$$

$$\frac{1}{9} < x$$

27. $\cosh x = 2$

$$\frac{e^x + e^{-x}}{2} = 2$$

$$e^x + e^{-x} = 4$$

$$e^x + \frac{1}{e^x} = 4$$

$$e^{2x} + 1 = 4e^x$$

$$e^{2x} - 4e^x + 1 = 0$$

$$e^x = \frac{4 \pm \sqrt{16-4}}{2} = 2 \pm \sqrt{3}$$

$$x = \ln (2+\sqrt{3})$$

$$x = \ln (2-\sqrt{3})$$

31. $A(t) = A_0 e^{-kt}$

$$\frac{A_0}{2} = A_0 e^{-k(28)}$$

$$\frac{1}{2} = e^{-28k}$$

$$\ln \frac{1}{2} = \ln e^{-28k} = -28k \,(\ln\ e)$$

$$\ln \frac{1}{2} = -28k$$

$$k = -\frac{1}{28} \ln \frac{1}{2} = .02476$$

33. $P = 50e^{-t/250}$

$$25 = 50e^{-t/250}$$

$$.5 = e^{-t/250}$$

$$\ln .5 = \ln e^{-t/250}$$

$$\ln .5 = \frac{-t}{250}$$

$$t = -250 \ln .5 = 173.3$$

35. $P = P_0 e^{kt}$

$$2P_0 = P_0 e^{.0132t}$$

$$2 = e^{.0132t}$$

$$\ln 2 = \ln e^{.0132t}$$

$$\ln 2 = .0132t$$

$$t = \frac{1}{.0132} \ln 2 = 52.5$$

Chapter 5   Review

1. $2^{x-5} = 3^x$

$$\ln 2^{x-5} = \ln 3^x$$

$$(x-5) \ln 2 = x \ln 3$$

$$x(\ln 2 - \ln 3) = 5 \ln 2$$

$$x = \frac{5 \ln 2}{\ln 2 - \ln 3} = -8.55$$

3. $2^{x^2-3x} = 16$

$$2^{x^2-3x} = 2^4$$

$$x^2 - 3x = 4$$

$$x^2 - 3x - 4 = 0$$

$$(x-4)(x+1) = 0$$

$$x = -1, 4$$

5.
$$3^{1-x} = 5$$
$$\ln 3^{1-x} = \ln 5$$
$$(1-x)\ln 3 = \ln 5$$
$$x \ln 3 = \ln 3 - \ln 5$$
$$x = \frac{\ln 3 - \ln 5}{\ln 3} = -.465$$

7. $\log x = 3,$
$$x = 10^3$$

9. $\log_3 81 = x$
$$3^x = 81 = 3^4$$
$$x = 4$$

11. $\log_x 64 = 6$
$$x^6 = 64$$
$$x^6 = 2^6$$
$$x = 2$$

13.
$$x^{4.3} = 2.1$$
$$\ln x^{4.3} = \ln 2.1$$
$$4.3 \ln x = \ln 2.1$$
$$\ln x = \frac{1}{4.3} \ln 2.1$$
$$x = 1.188$$

25. $\log x + \log x^2 = \log x^3 = 3 \log x$

27. $\log_2 2x + 3 \log_2 x = \log_2 2x + \log_2 x^3 = \log_2 2x^4$

29. $3 \log_5 x - \log_5 (2x-3) = \log_5 x^3 - \log_5 (2x-3) = \log_5 \frac{x^3}{2x-3}$

31.
$$2^{x+2} = 10$$
$$\log 2^{x+2} = \log 10$$
$$(x+2) \log 2 = 1$$
$$x \log 2 = 1 - 2 \log 2$$
$$x = \frac{1 - 2 \log 2}{\log 2} = 1.322$$

33.
$$3^{x+1} = 4^x$$
$$\ln 3^{x+1} = \ln 4^x$$
$$(x+1) \ln 3 = x \ln 4$$
$$x (\ln 3 - \ln 4) = -\ln 3$$
$$x = \frac{-\ln 3}{\ln 3 - \ln 4} = 3.819$$

35. $\log (\log x) = 1$
$$10^1 = \log x$$
$$10^{10} = \log x$$

37. $\log x + \log (x+1) = \log 6$
$$\log x(x+1) = \log 6$$
$$x^2 + x = 6$$
$$x^2 + x - 6 = 0$$
$$(x+3)(x-2) = 0$$
$$x = 2, \ x = \cancel{-3}, \ \log (-3) \text{ is not defined}$$

39. $\log x + \log (x-3) = 1$

$\qquad \log x(x-3) = 1$

$\qquad\qquad x^2-3x = 10$

$\qquad x^2-3x-10 = 0$

$\qquad (x-5)(x+2) = 0$

$\qquad\qquad x = 5, \cancel{-2}, \log (-2)$ is not defined

41. $A(t) = A_0 e^{-t/4}$

$\qquad \dfrac{A_0}{2} = A_0 e^{-t/4}$

$\qquad \ln \dfrac{1}{2} = \ln e^{-t/4}$

$\qquad \ln \dfrac{1}{2} = \dfrac{-t}{4}$

$\qquad t = -4 \ln \dfrac{1}{2} = 2.77$

45. $D(t) = D(0)e^{-rt/V}$

$\qquad r = \dfrac{1.4\text{L}}{\text{min}}, \; V = 450\text{mL} = .45\text{L}$

$\qquad t = 5$ sec. $= .0833$ min, $D(0) = 2.3$mg

$\qquad D(.0833) = 2.3e^{\frac{-1.4(.0833)}{.45}}$

$\qquad\qquad = 1.77$

- 69 -

Chapter 6 Systems of Equations and Inequalities

Section 6.1 Linear Systems of Equations

1. $\begin{aligned} 2x - 4y &= 2 \\ -2x + y &= 4 \end{aligned}$

$\qquad -3y = 6, \quad y = -2$

$\qquad -2x - 2 = 4, \quad x = -3$

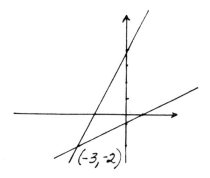

$(-3, -2)$

3. $\begin{aligned} 3x + 4y &= 23 \\ x - 3y &= -1 \end{aligned}$

$\qquad \begin{aligned} 3x + 4y &= 23 \\ -3x + 9y &= +3 \end{aligned}$

$\qquad 13y = 26, \quad y = 2$

$\qquad x - 6 = -1, \quad x = 5$

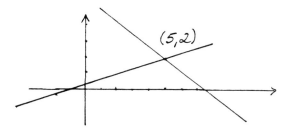

$(5, 2)$

5. $\begin{aligned} x - 4y &= 8 \\ 2x - 8y &= 16 \end{aligned}$

$\qquad \begin{aligned} -2x + 8y &= -16 \\ 2x - 8y &= 16 \end{aligned}$

$\qquad 0 = 0, \quad \text{dependent}$

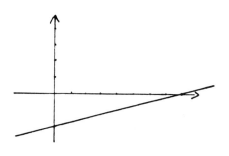

7. $\quad 3x + 4y = \quad 10$
$\quad\quad \underline{6x + 8y = -\;\; 2}$

$\quad\quad -6x - 8y = -20$
$\quad\quad \underline{\;\;\;6x + 8y = -\;\; 2}$
$\quad\quad\quad\quad\quad\; 0 \;= -22,\quad$ inconsistent

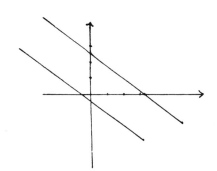

9. $\quad 3x - y = 2$
$\quad\quad \underline{2x + y = 6}$
$\quad\quad 5x \quad\quad = 8, \quad x = \frac{8}{5}$

$\quad\quad\quad 2\left(\frac{8}{5}\right) + y = 6, \quad y = \frac{14}{5}$

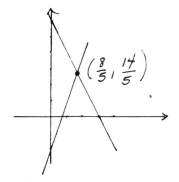

11. $\quad 3x + \;\; y = 0$
$\quad\quad \underline{2x - 2y = 2}$

$\quad\quad 6x + 2y = 0$
$\quad\quad \underline{2x - 2y = 2}$
$\quad\quad 8x \quad\quad = 2, \quad x = \frac{1}{4}$

$\quad\quad\quad 3\left(\frac{1}{4}\right) + y = 0, \quad y = \frac{-3}{4}$

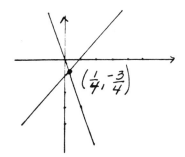

13. $\quad x + 7y - 7 = \;\; 0$
$\quad\quad \underline{8x - 7y - 3 = \;\; 0}$
$\quad\quad 9x \quad\quad\quad\quad = 10, \quad x = \frac{10}{9}$

$\quad\quad\quad \frac{10}{9} + 7y - 7 = 0, \quad y = \frac{53}{63}$

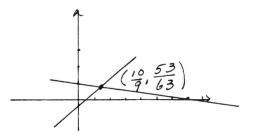

15. $r - s = 3$
    $\underline{r - 2s = 5}$
    $r = 3 + s$
    $(3+s) - 2s = 5, \quad s = -2$
    $r = 3-2 = 1$

17. $x + 3y = 7$
    $\underline{2x + 6y = 9}$
    $x = 7 - 3y$
    $2(7-3y) + 6y = 9$
    $\qquad 14 = 9,$ inconsistent

19. $5y - x = 2$
    $\underline{2y + 3x = 11}$
    $x = 5y-2,$
    $\qquad 2y + 3(5y-2) = 11, \ y=1$
    $\qquad\qquad x = 5(1) -2 = 3$

21. $4x + 5y = -6$
    $\underline{3x - 2y = -16}$
    $x = \dfrac{-6 - 5y}{4}$
    $\qquad 3\left(\dfrac{-6-5y}{4}\right) - 2y = -16,$
    $\qquad -18-15y-2y(4) = 64, \ y=2$
    $x = \dfrac{-6-5(2)}{4} = -4$

23. $6x - 8y = 14$
    $\underline{3x - 4y = 7}$
    $x = \dfrac{7 + 4y}{3}$
    $\qquad 6\left(\dfrac{7+4y}{3}\right) - 8y = 14$
    $14 + 8y - 8y = 14$
    dependent

25. $3x + 5y = 5$
    $\underline{x + 4y = 11}$
    $x = 11 - 4y$
    $\qquad 3(11-4y) + 5y = 5, \ y=4$
    $x = 11 - 4(4) = -5$

27. $9z - 13w = -3$
    $\underline{6z - 7w = 3}$
    $z = \dfrac{7w + 3}{6}$
    $9\left(\dfrac{7w+3}{6}\right) - 13w = -3$
    $9 + 21w - 26w = -6, \ w=3$
    $z = \dfrac{7(3) + 3}{6} = 4$

29. $2x + 3y = -10$
    $\underline{3x - 2y = -2}$
    $x = \dfrac{-10 - 3y}{2}$
    $3\left(\dfrac{-10 - 3y}{2}\right) - 2y = -2$
    $-30 - 9y - 4y = -4, \ y = +-2$
    $x = \dfrac{-10 - 3(-2)}{2} = -2$

31. $A + B = 25$
    $\underline{A - B = 3}$
    $2A = 28, \quad A = 14$
    $14 + B = 25, \quad B = 18$

33. $x + y = 12$

$\underline{x - y = 4}$

$2x = 16, \quad x = 8$

$\qquad 8 + y = 12, \; y = 4$

35. $ax + by^2 = 2$

$(4, -2) \rightarrow \qquad 4a + 4b = 2$

$(-2, 2) \rightarrow \qquad \underline{-2a + 4b = 2}$

$\qquad\qquad 6a = 0, \quad a = 0$

$\qquad\qquad 4b = 2, \quad b = \frac{1}{2}$

37. rate $\times$ time $=$ distance

$(v - w)\, 10 = 60$

$\underline{(v + w) 8 = 60}$

$80v - 80w = 480$

$\underline{80v + 80w = 600}$

$\qquad\quad 160v = 1080$

$\qquad\qquad\; v = 6.75$

$67.5 - 10w = 60, \quad w = .75$

39.

	Lead	Zinc
A_1	.80	.20
A_2	.50	.50
$A_1 + A_2$	.60	.40

$.80 A_1 + .50 A_2 = .60 \,(10)$

$.20 A_1 + .50 A_2 = .40 \,(10)$

$.60 A_1 = 2, \quad A_1 = \dfrac{10}{3}$

$A_1 + A_2 = 10, \quad A_2 = 10 - \dfrac{10}{3} = \dfrac{20}{3}$

41. $n + d = 35$

$\underline{.05n + .10d = 2.50}$

$10n + 10d = 350$

$\underline{5n + 10d = 250}$

$5n = 100, \; n = 20$

$d = 35 - n$

$= 35 - 20$

$\boxed{= 15}$ *nickels*

20 nickels

43. $C_1 = x$ min to do job

$C_2 = y$ min to do job

$\dfrac{1}{x} =$ amt done by C_1 in 1 min

$\dfrac{1}{y} =$ amt done by C_2 in 1 min

$10\left(\dfrac{1}{x} + \dfrac{1}{y}\right) = 1$

$6\left(\dfrac{1}{x} + \dfrac{1}{y}\right) + 9\left(\dfrac{1}{y}\right) = 1$

$\dfrac{10}{x} + \dfrac{10}{y} = 1$

$\dfrac{6}{x} + \dfrac{15}{y} = 1$

$\overline{}$

$\dfrac{60}{x} + \dfrac{60}{y} = 6$

$-\dfrac{60}{x} - \dfrac{150}{y} = -10$

$\overline{}$

$-\dfrac{90}{y} = -\,4, \; y = 22.5$ min

$\dfrac{10}{x} + \dfrac{10}{22.5} = 1$

$10(22.5) + 10x = 22.5$

$x = 18$ min

45. $\dfrac{x+550}{y} = \tan 29.1°$

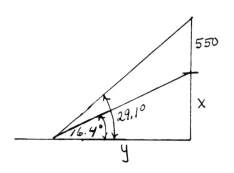

$\dfrac{x}{y} = \tan 16.4°, \ x = y \tan 16.4°$

$y \tan 16.4° + 550 = y \tan 29.1°$

$y = \dfrac{-550}{\tan 16.4° - \tan 29.1°} = 2097$

$x = 2097 \tan 16.4° = 617.2$

47. Demand: $\begin{aligned} x &= 100 \quad p = 400 \\ x &= 150 \quad p = 250 \end{aligned}$

$p - 400 = \left(\dfrac{400 - 250}{100 - 150}\right)(x - 100)$

$p = -3x + 700$

Supply: $\begin{aligned} x &= \ \ 75 \quad p = 300 \\ x &= 150 \quad p = 350 \end{aligned}$

$p - 300 = \left(\dfrac{350 - 300}{150 - 75}\right)(x - 75)$

$p = \dfrac{2x}{3} + 250 \quad 225$

$-3x + 700 = \dfrac{2x}{3} + 250$

$x = 123, \ p = -3(123) + 700 = 331$

Section 6.2   Nonlinear Systems

1.

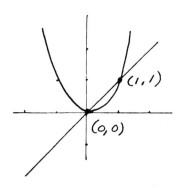

$(1, 1)$

$(0, 0)$

3.

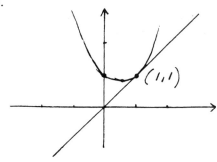

$(1, 1)$

5.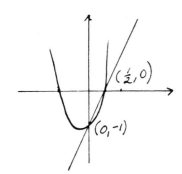

$(\frac{1}{2}, 0)$

$(0, -1)$

7.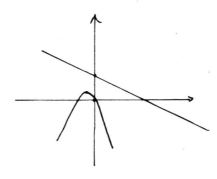

9.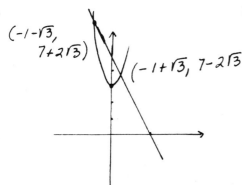

$(-1-\sqrt{3},\ 7+2\sqrt{3})$

$(-1+\sqrt{3},\ 7-2\sqrt{3})$

11.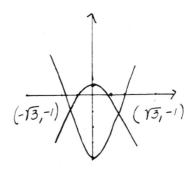

$(-\sqrt{3}, -1)$

$(\sqrt{3}, -1)$

13.
$$\begin{array}{rl} x^2 + y^2 &= 4 \\ 4x^2 + y^2 &= 4 \\ \hline -3x^2 &= 0 \\ x &= 0 \end{array}$$

$$y^2 = 4, \quad y = \pm 2$$

$$(0, 2), \quad (0, -2)$$

15.
$$\begin{array}{rl} x^2 + 4y^2 &= 2 \\ x^2 - y^2 &= 3 \\ \hline 5y^2 &= -1 \end{array}$$

No solution

17.
$$\begin{array}{rl} x^2 + y^2 &= 1 \\ 2x^2 - y^2 &= 2 \\ \hline 3x^2 &= 3 \end{array}$$

$$x^2 = 1, \quad x = \pm 1$$

$$y^2 = 0,$$

$$(1, 0), \quad (-1, 0)$$

19.
$$\begin{array}{rl} y - x &= 1 \\ x^2 - y^2 &= 10 \end{array}$$

$$y = 1 + x$$

$$x^2 - (1+x)^2 = 10$$

$$x^2 - 1 - 2x - x^2 = 10$$

$$x = \frac{-11}{2}$$

$$y = 1 - \frac{11}{2} = \frac{-9}{2}$$

$$\left(-\frac{11}{2}, \frac{-9}{2}\right)$$

21. $y^2 + 2x^2 = 2$
$\underline{y^2 + x^2 = 2}$

$x^2 = 0, \quad x = 0$
$y^2 = 2, \quad y = \pm\sqrt{2}$

$(0, \sqrt{2}), \quad (0, -\sqrt{2})$

23. $2a^2 - 4b^2 = 7$
$\underline{a^2 + 2b^2 = 3}$

$2a^2 - 4b^2 = 7$
$\underline{2a^2 - 4b^2 = 6}$

$4a^2 = 13, \quad a^2 = \dfrac{13}{4}$
$2b^2 = 3 - a^2 = 3 - \dfrac{13}{4}$
$b^2 = \dfrac{-1}{8}$, no solution

25. $ab = 16$
$\underline{a - b = 0}$

$a = b$
$b^2 = 16, \quad b = \pm 4$

$a = \pm 4$
$(4, 4), \quad (-4, -4)$

27. $xy = 1$
$\underline{x^2 + y^2 = 1}$

$x = \dfrac{1}{y}$

$\dfrac{1}{y^2} + y^2 = 1$

$1 + y^4 = y^2$

$y^4 - y^2 + 1 = 0$

$y^4 = \dfrac{1 \pm \sqrt{1 - 4}}{2}$

No real solutions

29. $x^2 + y^2 = 16$
$y = 2$
$x^2 + 4 = 16$
$x^2 = 12, \quad x = \pm\sqrt{12}$
$(2\sqrt{3}, 2), \quad (-2\sqrt{3}, 2)$

31. $x + y = 1$
$\underline{x^2 - y^2 = 1}$

$x = 1 - y$
$(1-y)^2 - y^2 = 1$
$1 - 2y + y^2 - y^2 = 1$
$y = 0$
$x = 1, \quad (1, 0)$

33. $x + y = 30$
$\underline{xy = 225}$

$x = 30 - y$
$(30 - y)y = 225$
$30y - y^2 - 225 = 0$
$(y - 15)(y - 15) = 0$
$y = 15, \ x = 15$

35. $p = x + 6$
$p = \dfrac{208}{x + 3}$

$x + 6 = \dfrac{208}{x + 3}$

$(x + 6)(x + 3) = 208$
$x^2 + 9x + 18 - 208 = 0$
$x^2 + 9x - 190 = 0$
$(x - 10)(x + 19) = 0$
$x = 10, \ \cancel{-19}$
$p = 10 + 6 = 16$

1. $\begin{bmatrix} 1 & 2 & | & 10 \\ 1 & -2 & | & 6 \end{bmatrix} \xrightarrow{-①+②} \begin{bmatrix} 1 & 2 & | & 10 \\ 0 & -4 & | & -16 \end{bmatrix} \xrightarrow{-\frac{1}{4}②} \begin{bmatrix} 1 & 2 & | & 10 \\ 0 & 1 & | & 4 \end{bmatrix}$

$x + 2y = 10$

$y = 4, \ x = 10 - 2(4) = 2$

3. $\begin{bmatrix} 3 & -1 & | & 5 \\ 1 & 4 & | & 6 \end{bmatrix} \xrightarrow{\frac{1}{3}①} \begin{bmatrix} 1 & -\frac{1}{3} & | & \frac{5}{3} \\ 1 & 4 & | & 6 \end{bmatrix} \xrightarrow{-①+②} \begin{bmatrix} 1 & -\frac{1}{3} & | & \frac{5}{3} \\ 0 & \frac{13}{3} & | & \frac{13}{3} \end{bmatrix}$

$x - \frac{1}{3}y = \frac{5}{3}$

$y = 1, \ x = \frac{5}{3} + \frac{1}{3} = 2$

5. $\begin{bmatrix} 8 & 5 & 0 & | & 4 \\ 0 & 5 & -3 & | & 1 \\ 12 & 0 & 5 & | & 6 \end{bmatrix} \xrightarrow{\frac{1}{8}①} \begin{bmatrix} 1 & \frac{5}{8} & 0 & | & \frac{1}{2} \\ 0 & 5 & -3 & | & 1 \\ 12 & 0 & 5 & | & 6 \end{bmatrix} \xrightarrow{-12①+③}$

$\begin{bmatrix} 1 & \frac{5}{8} & 0 & | & \frac{1}{2} \\ 0 & 5 & -3 & | & 1 \\ 0 & -\frac{15}{2} & 5 & | & 0 \end{bmatrix} \xrightarrow{\frac{1}{5}②} \begin{bmatrix} 1 & \frac{5}{8} & 0 & | & \frac{1}{2} \\ 0 & 1 & -\frac{3}{5} & | & \frac{1}{5} \\ 0 & -\frac{15}{2} & 5 & | & 0 \end{bmatrix} \begin{array}{l} \xrightarrow{\frac{15}{2}②+③} \\ \xrightarrow{-\frac{5}{8}②+①} \end{array}$

$\begin{bmatrix} 1 & 0 & \frac{3}{8} & | & \frac{3}{8} \\ 0 & 1 & -\frac{3}{5} & | & \frac{1}{5} \\ 0 & 0 & \frac{1}{2} & | & \frac{3}{2} \end{bmatrix}$

$x + \frac{3}{8}z = \frac{3}{8}$

$y - \frac{3z}{5} = \frac{1}{5}$

$\frac{1}{2}z = \frac{3}{2}, \ z = 3$

$y = \frac{1}{5} + \frac{3}{5}(3) = 2$

$x = \frac{3}{8} - \frac{3}{8}(3) = -\frac{3}{4}$

7.
$$
\begin{bmatrix} 3 & 5 & -3 & | & 31 \\ 2 & -3 & 2 & | & 13 \\ 5 & 2 & -5 & | & 20 \end{bmatrix}
\xrightarrow{\;①+③\;}
\begin{bmatrix} 3 & 5 & --3 & | & 31 \\ 2 & -3 & 2 & | & 13 \\ 2 & -3 & -2 & | & -11 \end{bmatrix}
\xrightarrow[\;-②+③\;]{\;-②+①\;}
$$

$$
\begin{bmatrix} 1 & 8 & -5 & | & 18 \\ 2 & -3 & 2 & | & 13 \\ 0 & 0 & -4 & | & -24 \end{bmatrix}
\xrightarrow[\;2①+②\;]{\;-\frac{1}{4}③\;}
\begin{bmatrix} 1 & 8 & -5 & | & 18 \\ 0 & -19 & 12 & | & -23 \\ 0 & 0 & 1 & | & 6 \end{bmatrix}
$$

$$
x + 8y - 5z = 18
$$
$$
-19y + 12z = -23
$$
$$
z = 6
$$
$$
-19y = -23 - 12(6), \quad y = 5
$$
$$
x = -8(5) + 5(6) + 18 = 8
$$

9.
$$
\begin{bmatrix} 2 & 0 & -5 & | & 2 \\ 0 & 9 & -4 & | & 7 \\ 3 & -12 & 0 & | & -2 \end{bmatrix}
\xrightarrow{\;-①+③\;}
\begin{bmatrix} 2 & 0 & -5 & | & 2 \\ 0 & 9 & -4 & | & 7 \\ 1 & -12 & 5 & | & -4 \end{bmatrix}
\xrightarrow{\;-③+①\;}
$$

$$
\begin{bmatrix} 1 & 12 & -10 & | & 6 \\ 0 & 9 & -4 & | & 7 \\ 1 & -12 & 5 & | & 4 \end{bmatrix}
\xrightarrow{\;-①+③\;}
\begin{bmatrix} 1 & 12 & -10 & | & 6 \\ 0 & 9 & -4 & | & 7 \\ 0 & -24 & 15 & | & -10 \end{bmatrix}
\xrightarrow{\;\frac{1}{9}②\;}
$$

$$
\begin{bmatrix} 1 & 12 & -10 & | & 6 \\ 0 & 1 & -\frac{4}{9} & | & \frac{7}{9} \\ 0 & -24 & 15 & | & -10 \end{bmatrix}
\xrightarrow{\;24②+③\;}
\begin{bmatrix} 1 & 12 & -10 & | & 6 \\ 0 & 1 & -\frac{4}{9} & | & \frac{7}{9} \\ 0 & 0 & \frac{13}{3} & | & \frac{26}{3} \end{bmatrix}
$$

$$
x + 12y - 10z = 6
$$
$$
y - \frac{4}{9}z = \frac{7}{9}
$$
$$
\frac{13z}{3} = \frac{26}{3}, \; z = 2
$$
$$
y = \frac{7}{9} + \frac{4}{9}(2) = \frac{5}{3}
$$
$$
x = 6 - 12\left(\frac{5}{3}\right) + 10(2) = 6
$$

11.
$$\begin{bmatrix} 1 & 1 & 2 & | & 3 \\ 3 & -1 & -1 & | & 1 \\ 1 & 5 & 9 & | & 11 \end{bmatrix} \xrightarrow[-①+③]{-3①+②} \begin{bmatrix} 1 & 1 & 2 & | & 3 \\ 0 & -4 & -7 & | & -8 \\ 0 & 4 & 7 & | & 8 \end{bmatrix} \xrightarrow{②+③}$$

$$\begin{bmatrix} 1 & 1 & 2 & | & 3 \\ 0 & -4 & -7 & | & 8 \\ 0 & 0 & 0 & | & 0 \end{bmatrix} \qquad \text{Dependent}$$

13.
$$\begin{bmatrix} 2 & 1 & -1 & | & 9 \\ 1 & -1 & 1 & | & 0 \\ -1 & 3 & -2 & | & 5 \end{bmatrix} \xrightarrow{③+①} \begin{bmatrix} 1 & 4 & 3 & | & 14 \\ 1 & -1 & 1 & | & 0 \\ -1 & 3 & -2 & | & 5 \end{bmatrix} \xrightarrow{②+③}$$

$$\begin{bmatrix} 1 & 4 & -3 & | & 14 \\ 1 & -1 & 1 & | & 0 \\ 0 & 2 & -1 & | & 5 \end{bmatrix} \xrightarrow{-①+②} \begin{bmatrix} 1 & 4 & -3 & | & 14 \\ 0 & -5 & 4 & | & -14 \\ 0 & 2 & -1 & | & 5 \end{bmatrix} \xrightarrow[\cdot\frac{1}{2}③]{\frac{-1}{5}②}$$

$$\begin{bmatrix} 1 & 4 & -3 & | & 14 \\ 0 & 1 & -\frac{4}{5} & | & \frac{14}{5} \\ 0 & 1 & -\frac{1}{2} & | & \frac{5}{2} \end{bmatrix} \xrightarrow{-②+③} \begin{bmatrix} 1 & 4 & -3 & | & 14 \\ 0 & 1 & -\frac{4}{5} & | & \frac{14}{5} \\ 0 & 0 & \frac{3}{10} & | & -\frac{3}{10} \end{bmatrix}$$

$$x + 4y - 3z = 14$$
$$y - \frac{4}{5}z = \frac{14}{5}$$
$$\frac{3z}{10} = \frac{-3}{10}, \; z = -1$$
$$y = \frac{14}{5} + \frac{4}{5}(-1) = 2$$
$$x = 14 - 4(2) + 3(-1) = 3$$

15.
$$\begin{bmatrix} 1 & -1 & 0 & | & 2 \\ 0 & 1 & -1 & | & -3 \\ 3 & 2 & -5 & | & 1 \end{bmatrix} \xrightarrow{-3①+③} \begin{bmatrix} 1 & -1 & 0 & | & 2 \\ 0 & 1 & -1 & | & -3 \\ 0 & 5 & -5 & | & -5 \end{bmatrix} \xrightarrow{-5②+③}$$

$$\begin{bmatrix} 1 & -1 & 0 & | & 2 \\ 0 & 1 & -1 & | & -3 \\ 0 & 0 & 0 & | & 10 \end{bmatrix} \qquad \text{Inconsistent}$$

- 79 -

17.

$$\begin{bmatrix} 1 & -3 & 2 & | & 0 \\ 2 & 1 & -1 & | & -3 \\ 10 & -2 & 0 & | & -12 \end{bmatrix} \xrightarrow[\;-10\,①+③\;]{-2\,①+②} \begin{bmatrix} 1 & -3 & 2 & | & 0 \\ 0 & 7 & -5 & | & -3 \\ 0 & 28 & -20 & | & -12 \end{bmatrix} \xrightarrow{-4\,②+③}$$

$$\begin{bmatrix} 1 & -3 & 2 & | & 0 \\ 0 & 7 & 5 & | & -3 \\ 0 & 0 & 0 & | & 0 \end{bmatrix} \qquad \text{Dependent}$$

19.

$$\begin{bmatrix} 1 & 1 & 1 & | & 3 \\ 1 & -2 & -3 & | & 4 \\ 2 & 1 & 2 & | & 9 \end{bmatrix} \xrightarrow[\;-2\,①+③\;]{-①+②} \begin{bmatrix} 1 & 1 & 1 & | & 3 \\ 0 & -3 & -4 & | & 1 \\ 0 & -1 & 0 & | & 3 \end{bmatrix}$$

$$x + y + z = 3$$
$$-3y - 4z = 1$$
$$-y = 3, \quad y = -3$$
$$4z = -1 - 3(-3), \; z = 2$$
$$x = 3 - (-3) - (2) = 4$$

21.

$$\begin{bmatrix} 3 & -2 & 1 & 1 & | & 10 \\ 1 & 1 & -1 & -1 & | & -5 \\ 2 & 1 & 1 & 5 & | & 18 \\ -1 & -2 & 5 & -2 & | & 5 \end{bmatrix} \xrightarrow[\substack{2\,④+③ \\ 3\,④+①}]{④+②} \begin{bmatrix} 0 & -8 & 16 & -5 & | & 25 \\ 0 & -1 & 4 & -3 & | & 0 \\ 0 & -3 & 11 & 1 & | & 28 \\ -1 & -2 & 5 & -2 & | & 5 \end{bmatrix} \xrightarrow[\substack{④\leftrightarrow① \\ -②}]{-④}$$

$$\begin{bmatrix} 1 & 2 & -5 & 2 & | & -5 \\ 0 & 1 & -4 & 3 & | & 0 \\ 0 & -3 & 11 & 1 & | & 28 \\ 0 & -8 & 16 & -5 & | & 25 \end{bmatrix} \xrightarrow[\substack{3\,②+③ \\ 8\,②+④}]{-2\,②+①} \begin{bmatrix} 1 & 0 & 3 & -4 & | & -5 \\ 0 & 1 & -4 & 3 & | & 0 \\ 0 & 0 & -1 & 10 & | & 28 \\ 0 & 0 & -16 & 19 & | & 25 \end{bmatrix} \xrightarrow[\substack{-4\,③+② \\ -16\,③+④}]{3\,③+①}$$

$$\begin{bmatrix} 1 & 0 & 0 & 26 & | & 79 \\ 0 & 1 & 0 & -37 & | & -112 \\ 0 & 0 & -1 & 10 & | & 28 \\ 0 & 0 & 0 & -141 & | & -423 \end{bmatrix}$$

$$x + 26w = 79$$
$$y - 37w = -112$$
$$-z + 10w = 28$$
$$-141w = -423, \; w = 3$$
$$z = -28 + 10(3) = 2$$
$$y = -112 + 37(3) = -1$$
$$x = 79 - 26(3) = 1$$

23.

$$
\left[\begin{array}{cccc|c}
3 & 1 & 1 & -1 & -2 \\
-5 & 2 & -3 & -3 & -1 \\
7 & 5 & 4 & -1 & -1 \\
4 & -1 & -2 & 3 & 7
\end{array}\right]
\qquad
\left[\begin{array}{cccc|c}
3 & 1 & 1 & -1 & -2 \\
-5 & 2 & -3 & -3 & -1 \\
7 & 5 & 4 & -1 & -1 \\
7 & 0 & -1 & 2 & 5
\end{array}\right]
$$

$$
\left[\begin{array}{cccc|c}
3 & 1 & 1 & -1 & -2 \\
1 & 4 & -1 & -5 & -5 \\
0 & 5 & 5 & -3 & -6 \\
7 & 0 & -1 & 2 & 5
\end{array}\right]
\qquad
\left[\begin{array}{cccc|c}
0 & -11 & 4 & 14 & 13 \\
1 & 4 & -1 & -5 & -5 \\
0 & 5 & 5 & -3 & -6 \\
0 & -28 & 6 & 37 & 40
\end{array}\right]
$$

$$
\left[\begin{array}{cccc|c}
0 & -11 & 4 & 14 & 13 \\
1 & 4 & -1 & -5 & -5 \\
0 & 5 & 5 & -3 & -6 \\
0 & 5 & -6 & -5 & 1
\end{array}\right]
\qquad
\left[\begin{array}{cccc|c}
0 & -1 & 14 & 8 & 1 \\
1 & 4 & -1 & -5 & -5 \\
0 & 5 & 5 & -3 & -6 \\
0 & 0 & -11 & -2 & 7
\end{array}\right]
$$

$$
\left[\begin{array}{cccc|c}
0 & -1 & 14 & 8 & 1 \\
1 & 4 & -1 & -5 & -5 \\
0 & 0 & 75 & 37 & -1 \\
0 & 0 & -11 & -2 & 7
\end{array}\right]
\qquad
\left[\begin{array}{cccc|c}
0 & 1 & -14 & -8 & -1 \\
1 & 4 & -1 & -5 & -5 \\
0 & 0 & -2 & 23 & 48 \\
0 & 0 & -11 & -2 & 7
\end{array}\right]
$$

$$
\left[\begin{array}{cccc|c}
1 & 4 & -1 & -5 & -5 \\
0 & 1 & -14 & -8 & -1 \\
0 & 0 & -2 & 23 & 48 \\
0 & 0 & -1 & -117 & -233
\end{array}\right]
\qquad
\left[\begin{array}{cccc|c}
1 & 4 & -1 & -5 & -5 \\
0 & 1 & -14 & -8 & -1 \\
0 & 0 & 0 & 257 & 514 \\
0 & 0 & -1 & -117 & -233
\end{array}\right]
$$

$$
\begin{aligned}
x + 4y - z - 5w &= -5 \\
y - 14z - 8w &= -1 \\
257w &= 514 \\
-z - 117w &= -233
\end{aligned}
$$

$$
\begin{aligned}
w &= 2 \\
z &= 233 - 117(2) = -1 \\
y &= -1 + 8(2) + 14(-1) = 1 \\
x &= -5 + 5(2) - (-1) - 4(1) = 1
\end{aligned}
$$

25. $n + d + q = 900$

$n = 20 + d, \qquad n - d = 20$

$n + d = 44 + q, \qquad n + d - q = 44$

$$
\left[\begin{array}{ccc|c}
1 & 1 & 1 & 900 \\
1 & -1 & 0 & 20 \\
1 & 1 & -1 & 44
\end{array}\right]
\begin{array}{l} -①+② \\ \longrightarrow \\ -①+③ \end{array}
\left[\begin{array}{ccc|c}
1 & 1 & 1 & 900 \\
0 & -2 & -1 & -880 \\
0 & 0 & -2 & -856
\end{array}\right]
\begin{array}{l} -\frac{1}{2}③ \\ \longrightarrow \\ -② \end{array}
$$

$$
\left[\begin{array}{ccc|c}
1 & 1 & 1 & 900 \\
0 & 2 & 1 & 880 \\
0 & 0 & 1 & 428
\end{array}\right]
\begin{array}{l} -③+② \\ \longrightarrow \\ -③+① \end{array}
\left[\begin{array}{ccc|c}
1 & 1 & 0 & 472 \\
0 & 2 & 0 & 452 \\
0 & 0 & 1 & 428
\end{array}\right]
\begin{array}{l} \frac{1}{2}② \\ \longrightarrow \end{array}
$$

$$
\left[\begin{array}{ccc|c}
1 & 1 & 0 & 472 \\
0 & 1 & 0 & 226 \\
0 & 0 & 1 & 428
\end{array}\right]
\begin{array}{l} -②+① \\ \longrightarrow \end{array}
\left[\begin{array}{ccc|c}
1 & 0 & 0 & 246 \\
0 & 1 & 0 & 226 \\
0 & 0 & 1 & 428
\end{array}\right]
$$

27. $l_1 + l_2 + l_3 = 45$

$2l_1 = l_2 + l_3, \qquad 2l_1 - l_2 - l_3 = 0$

$l_2 = 4 + l_3, \qquad l_2 - l_3 = 4$

$$
\left[\begin{array}{ccc|c}
1 & 1 & 1 & 45 \\
2 & -1 & -1 & 0 \\
0 & 1 & -1 & 4
\end{array}\right]
\begin{array}{l} -2①+② \\ \longrightarrow \end{array}
\left[\begin{array}{ccc|c}
1 & 1 & 1 & 45 \\
0 & -3 & -3 & -90 \\
0 & 1 & -1 & 4
\end{array}\right]
\begin{array}{l} -\frac{1}{3}② \\ \longrightarrow \end{array}
$$

$$
\left[\begin{array}{ccc|c}
1 & 1 & 1 & 45 \\
0 & 1 & 1 & 30 \\
0 & 1 & -1 & 4
\end{array}\right]
\begin{array}{l} -②+③ \\ \longrightarrow \end{array}
\left[\begin{array}{ccc|c}
1 & 1 & 1 & 45 \\
0 & 1 & 1 & 30 \\
0 & 0 & -2 & -26
\end{array}\right]
\begin{array}{l} -\frac{1}{2}③ \\ \longrightarrow \end{array}
$$

$$
\left[\begin{array}{ccc|c}
1 & 1 & 1 & 45 \\
0 & 1 & 1 & 30 \\
0 & 0 & 1 & 13
\end{array}\right]
\begin{array}{l} -③+② \\ \longrightarrow \\ -③+① \end{array}
\left[\begin{array}{ccc|c}
1 & 1 & 0 & 32 \\
0 & 1 & 0 & 17 \\
0 & 0 & 1 & 13
\end{array}\right]
\begin{array}{l} -②+① \\ \longrightarrow \end{array}
$$

$$
\left[\begin{array}{ccc|c}
1 & 0 & 0 & 15 \\
0 & 1 & 0 & 17 \\
0 & 0 & 1 & 13
\end{array}\right]
$$

1. $A + B = \begin{bmatrix} 5 & 0 \\ 1 & 1 \end{bmatrix} + \begin{bmatrix} -1 & 2 \\ 1 & 0 \end{bmatrix} = \begin{bmatrix} 4 & 2 \\ 2 & 1 \end{bmatrix}$

$A + B = \begin{bmatrix} 5 & 0 \\ 1 & 1 \end{bmatrix} - \begin{bmatrix} -1 & 2 \\ 1 & 0 \end{bmatrix} = \begin{bmatrix} 6 & -2 \\ 0 & 1 \end{bmatrix}$

$2A = 2\begin{bmatrix} 5 & 0 \\ 1 & 1 \end{bmatrix} = \begin{bmatrix} 10 & 0 \\ 2 & 2 \end{bmatrix}$

3. $A + B = \begin{bmatrix} 1 & 2 & 3 \\ 4 & 0 & 1 \end{bmatrix} + \begin{bmatrix} 1 & 0 & 2 \\ 0 & 5 & 6 \end{bmatrix} = \begin{bmatrix} 2 & 2 & 5 \\ 4 & 5 & 7 \end{bmatrix}$

$A - B = \begin{bmatrix} 1 & 2 & 3 \\ 4 & 0 & 1 \end{bmatrix} - \begin{bmatrix} 1 & 0 & 2 \\ 0 & 5 & 6 \end{bmatrix} = \begin{bmatrix} 0 & 2 & 1 \\ 4 & -5 & -5 \end{bmatrix}$

$2A = 2\begin{bmatrix} 1 & 2 & 3 \\ 4 & 0 & 1 \end{bmatrix} = \begin{bmatrix} 2 & 4 & 6 \\ 8 & 0 & 2 \end{bmatrix}$

5. $A + B = \begin{bmatrix} 1 \\ 0 \\ 2 \end{bmatrix} + \begin{bmatrix} 7 \\ -1 \\ 5 \end{bmatrix} = \begin{bmatrix} 8 \\ -1 \\ 7 \end{bmatrix}$

$A - B = \begin{bmatrix} 1 \\ 0 \\ 2 \end{bmatrix} + \begin{bmatrix} 7 \\ -1 \\ 5 \end{bmatrix} = \begin{bmatrix} -6 \\ 1 \\ -3 \end{bmatrix}$

$2A = 2\begin{bmatrix} 1 \\ 0 \\ 2 \end{bmatrix} = \begin{bmatrix} 2 \\ 0 \\ 4 \end{bmatrix}$

7. $A + B = \begin{bmatrix} 0 & 3 & 1 \end{bmatrix} + \begin{bmatrix} -1 & 2 & 0 \end{bmatrix} = \begin{bmatrix} -1 & 5 & 1 \end{bmatrix}$

$A - B = \begin{bmatrix} 0 & 3 & 1 \end{bmatrix} - \begin{bmatrix} -1 & 2 & 0 \end{bmatrix} = \begin{bmatrix} 1 & 1 & 1 \end{bmatrix}$

$2A = 2\begin{bmatrix} 0 & 3 & 1 \end{bmatrix} - \begin{bmatrix} 0 & 6 & 2 \end{bmatrix}$

11. $AB = \begin{bmatrix} 5 & 0 \\ 1 & 1 \end{bmatrix}\begin{bmatrix} -1 & 2 \\ 1 & 0 \end{bmatrix} = \begin{bmatrix} 5(-1)+0(1) & 5(2)+0(0) \\ 1(-1)+1(1) & 1(2)+1(0) \end{bmatrix} = \begin{bmatrix} -5 & 10 \\ 0 & 2 \end{bmatrix}$

$BA = \begin{bmatrix} -1 & 2 \\ 1 & 0 \end{bmatrix}\begin{bmatrix} -5 & 0 \\ 1 & 1 \end{bmatrix} = \begin{bmatrix} -1(5)+2(1) & -1(0)+2(1) \\ 1(5)+0(1) & 1(0)+0(1) \end{bmatrix} = \begin{bmatrix} -3 & 2 \\ 5 & 0 \end{bmatrix}$

15. $AB = \begin{bmatrix} -1 \\ 5 \end{bmatrix}\begin{bmatrix} 2 & 0 \end{bmatrix} = \begin{bmatrix} -2 & 0 \\ 10 & 0 \end{bmatrix}$

$BA = \begin{bmatrix} 2 & 0 \end{bmatrix}\begin{bmatrix} -1 \\ 5 \end{bmatrix} = \begin{bmatrix} 2(-1)+0(5) \end{bmatrix} = \begin{bmatrix} -2 \end{bmatrix}$

17. $AB = \begin{bmatrix} 2 & 1 \\ 1 & -1 \end{bmatrix}\begin{bmatrix} 1 \\ 2 \end{bmatrix} = \begin{bmatrix} 2(-1)+1(2) \\ 1(1)-1(2) \end{bmatrix} = \begin{bmatrix} 4 \\ -1 \end{bmatrix}$

BA not possible

19. $AB = \begin{bmatrix} 2 & 0 & 5 \\ 6 & 1 & 4 \\ -2 & 1 & 1 \end{bmatrix}\begin{bmatrix} 1 & 0 \\ -0 & 1 \\ 2 & 3 \end{bmatrix} =$

$\begin{bmatrix} 2(1)+0(-1)+5(2) & 2(0)+0(1)+5(3) \\ 6(1)+1(-1)+4(2) & 6(0)+1(1)+4(3) \\ -2(1)+1(-1)+(2) & -2(0)+1(1)+1(3) \end{bmatrix} = \begin{bmatrix} 12 & 15 \\ 13 & 13 \\ -1 & 4 \end{bmatrix}$

BA not possible

21. $AX = \begin{bmatrix} 2 & 1 \\ 1 & -1 \end{bmatrix} \begin{bmatrix} 1 \\ -1 \end{bmatrix} = \begin{bmatrix} 2(1)+1(-1) \\ 1(1)-1(-1) \end{bmatrix} = \begin{bmatrix} 1 \\ 2 \end{bmatrix} = B$

23. $AX = \begin{bmatrix} 2 & 1 & 1 \\ 1 & -1 & 1 \\ 1 & -2 & 2 \end{bmatrix} \begin{bmatrix} 1 \\ 1 \\ 3 \end{bmatrix} = \begin{bmatrix} 2(1)+1(1)+1(3) \\ 1(1)-1(1)+1(3) \\ 1(1)-2(1)+2(3) \end{bmatrix} = \begin{bmatrix} 6 \\ 3 \\ 5 \end{bmatrix} = B$

25. $AX = \begin{bmatrix} 2 & 1 & 3 \\ 1 & -2 & -1 \end{bmatrix} \begin{bmatrix} -\frac{1}{2} \\ 1 \\ 1 \end{bmatrix} = \begin{bmatrix} 2(-\frac{1}{2})+1(1)+3(1) \\ 1(-\frac{1}{2})-2(1)-1(1) \end{bmatrix} = \begin{bmatrix} 3 \\ -\frac{7}{2} \end{bmatrix} = B$

27. $BC = \begin{bmatrix} 0 & 1 \\ 5 & 2 \end{bmatrix} \begin{bmatrix} 3 & 5 \\ 0 & 2 \end{bmatrix} = \begin{bmatrix} 0 & 2 \\ 15 & 29 \end{bmatrix}$

$A(BC) = \begin{bmatrix} 1 & 2 \\ -1 & 0 \end{bmatrix} \begin{bmatrix} 0 & 2 \\ 15 & 29 \end{bmatrix} = \begin{bmatrix} 30 & 60 \\ 0 & -2 \end{bmatrix}$

$AB = \begin{bmatrix} 1 & 2 \\ -1 & 0 \end{bmatrix} \begin{bmatrix} 0 & 1 \\ 5 & 2 \end{bmatrix} = \begin{bmatrix} 10 & 5 \\ 0 & -1 \end{bmatrix}$

$(AB)C = \begin{bmatrix} 10 & 5 \\ 0 & -1 \end{bmatrix} \begin{bmatrix} 3 & 5 \\ 0 & 2 \end{bmatrix} = \begin{bmatrix} 30 & 60 \\ 0 & -2 \end{bmatrix}$

Thurs
get to SSC
11:59
Leave SSC
FRI: 1:50

29. $A^2 = \begin{bmatrix} 1 & 2 \\ -1 & 0 \end{bmatrix} \begin{bmatrix} 1 & 2 \\ -1 & 0 \end{bmatrix} = \begin{bmatrix} -1 & 2 \\ -1 & -2 \end{bmatrix}$

$B^2 = \begin{bmatrix} 0 & 1 \\ 5 & 2 \end{bmatrix} \begin{bmatrix} 0 & 1 \\ 5 & 2 \end{bmatrix} = \begin{bmatrix} 5 & 2 \\ 10 & 9 \end{bmatrix}$

$A^2 - B^2 = \begin{bmatrix} -1 & 2 \\ -1 & -2 \end{bmatrix} - \begin{bmatrix} 5 & 2 \\ 10 & 9 \end{bmatrix} = \begin{bmatrix} -6 & 0 \\ -11 & -11 \end{bmatrix}$

$A - B = \begin{bmatrix} 1 & 2 \\ -1 & 0 \end{bmatrix} - \begin{bmatrix} 0 & 1 \\ 5 & 2 \end{bmatrix} = \begin{bmatrix} 1 & 1 \\ -6 & -2 \end{bmatrix}$

$A + B = \begin{bmatrix} 1 & 2 \\ -1 & 0 \end{bmatrix} + \begin{bmatrix} 0 & 1 \\ 5 & 2 \end{bmatrix} = \begin{bmatrix} 1 & 3 \\ 4 & 2 \end{bmatrix}$

$(A-B)(A+B) = \begin{bmatrix} 1 & 1 \\ -6 & -2 \end{bmatrix} \begin{bmatrix} 1 & 3 \\ 4 & 2 \end{bmatrix} = \begin{bmatrix} 5 & 5 \\ -14 & -22 \end{bmatrix}$

31. $(A+B)^2 = (A+B)(A+B) = A(A+B) + B(A+B) = A^2 + AB + BA + B^2$

33. $AB = \begin{bmatrix} 0 & 1 \\ 0 & 1 \end{bmatrix} \begin{bmatrix} 1 & 1 \\ 1 & 0 \end{bmatrix} = \begin{bmatrix} 1 & 0 \\ 1 & 0 \end{bmatrix}$

$AC = \begin{bmatrix} 0 & 1 \\ 0 & 1 \end{bmatrix} \begin{bmatrix} 0 & 0 \\ 1 & 0 \end{bmatrix} = \begin{bmatrix} 1 & 0 \\ 1 & 0 \end{bmatrix}$

Section 6.5   Solving Systems of Equations by Matrix Inversion

1. $\left[\begin{array}{cc|cc} 0 & 1 & 1 & 0 \\ 1 & 0 & 0 & 1 \end{array} \right] \xrightarrow{①\leftrightarrow②} \left[\begin{array}{cc|cc} 1 & 0 & 0 & 1 \\ 0 & 1 & 1 & 0 \end{array} \right]$

3.
$$\left[\begin{array}{cc|cc} 2 & 1 & 1 & 0 \\ 1 & 0 & 0 & 1 \end{array}\right] \xrightarrow{\;①\leftrightarrow②\;} \left[\begin{array}{cc|cc} 1 & 0 & 0 & 1 \\ 2 & 1 & 1 & 0 \end{array}\right] \xrightarrow{\;-2①+②\;}$$

$$\left[\begin{array}{cc|cc} 1 & 0 & 0 & 1 \\ 0 & 1 & 1 & -2 \end{array}\right]$$

5.
$$\left[\begin{array}{ccc|ccc} 2 & 0 & 0 & 1 & 0 & 0 \\ 0 & 3 & 0 & 0 & 1 & 0 \\ 0 & 0 & 5 & 0 & 0 & 1 \end{array}\right] \begin{array}{c} \frac{1}{2}① \\ \frac{1}{3}② \\ \frac{1}{5}③ \end{array} \longrightarrow \left[\begin{array}{ccc|ccc} 1 & 0 & 0 & \frac{1}{2} & 0 & 0 \\ 0 & 1 & 0 & 0 & \frac{1}{3} & 0 \\ 0 & 0 & 1 & 0 & 0 & \frac{1}{5} \end{array}\right]$$

7.
$$\left[\begin{array}{ccc|ccc} 1 & 0 & 1 & 1 & 0 & 0 \\ 2 & 1 & 2 & 0 & 1 & 0 \\ -1 & 0 & 0 & 0 & 0 & 1 \end{array}\right] \begin{array}{c} ①+③ \\ -2①+② \end{array} \longrightarrow \left[\begin{array}{ccc|ccc} 1 & 0 & 1 & 1 & 0 & 0 \\ 0 & 1 & 0 & -2 & 1 & 0 \\ 0 & 0 & 1 & 1 & 0 & 1 \end{array}\right] \xrightarrow{\;-③+①\;}$$

$$\left[\begin{array}{ccc|ccc} 1 & 0 & 0 & 0 & 0 & -1 \\ 0 & 1 & 0 & -2 & 1 & 0 \\ 0 & 0 & 1 & 1 & 0 & 1 \end{array}\right]$$

11.
$$\left[\begin{array}{cc|cc} a & b & 1 & 0 \\ 0 & d & 0 & 1 \end{array}\right] \begin{array}{c} \frac{1}{a}① \\ \frac{1}{d}② \end{array} \longrightarrow \left[\begin{array}{cc|cc} 1 & \frac{b}{a} & \frac{1}{a} & 0 \\ 0 & 1 & 0 & \frac{1}{d} \end{array}\right] \xrightarrow{\;-\frac{b}{a}②+①\;} \left[\begin{array}{cc|cc} 1 & 0 & \frac{1}{a} & -\frac{b}{ad} \\ 0 & 1 & 0 & \frac{1}{d} \end{array}\right]$$

13.

$$\left[\begin{array}{ccc|ccc} 2 & 4 & 3 & 1 & 0 & 0 \\ -2 & 4 & -2 & 0 & 1 & 0 \\ 1 & 0 & 1 & 0 & 0 & 1 \end{array}\right] \xrightarrow{①+②} \left[\begin{array}{ccc|ccc} 2 & 4 & 3 & 1 & 0 & 0 \\ 0 & 8 & 1 & 1 & 1 & 0 \\ 1 & 0 & 1 & 0 & 0 & 1 \end{array}\right] \xrightarrow{-③+①}$$

$$\left[\begin{array}{ccc|ccc} 1 & 4 & 2 & 1 & 0 & -1 \\ 0 & 8 & 1 & 1 & 1 & 0 \\ 1 & 0 & 1 & 0 & 0 & 1 \end{array}\right] \xrightarrow{-①+③} \left[\begin{array}{ccc|ccc} 1 & 4 & 2 & 1 & 0 & -1 \\ 0 & 8 & 1 & 1 & 1 & 0 \\ 0 & -4 & -1 & -1 & 0 & 2 \end{array}\right] \xrightarrow[\,-\frac{1}{4}③\,]{\frac{1}{8}②}$$

$$\left[\begin{array}{ccc|ccc} 1 & 4 & 2 & 1 & 0 & -1 \\ 0 & 1 & \frac{1}{8} & \frac{1}{8} & \frac{1}{8} & 0 \\ 0 & 1 & \frac{1}{4} & \frac{1}{4} & 0 & -\frac{1}{2} \end{array}\right] \xrightarrow{-②+③} \left[\begin{array}{ccc|ccc} 1 & 4 & 2 & 1 & 0 & -1 \\ 0 & 1 & \frac{1}{8} & \frac{1}{8} & \frac{1}{8} & 0 \\ 0 & 0 & \frac{1}{8} & \frac{1}{8} & -\frac{1}{8} & -\frac{1}{2} \end{array}\right] \xrightarrow[-4②+①]{8③}$$

$$\left[\begin{array}{ccc|ccc} 1 & 0 & \frac{3}{2} & \frac{1}{2} & -\frac{1}{2} & -1 \\ 0 & 1 & \frac{1}{8} & \frac{1}{8} & \frac{1}{8} & 0 \\ 0 & 0 & 1 & 1 & -1 & -4 \end{array}\right] \xrightarrow[-\frac{1}{8}③+②]{-\frac{3}{2}③+①} \left[\begin{array}{ccc|ccc} 1 & 0 & 0 & -1 & 1 & 5 \\ 0 & 1 & 0 & 0 & \frac{1}{4} & \frac{1}{2} \\ 0 & 0 & 1 & 1 & -1 & -4 \end{array}\right]$$

15.

$$\left[\begin{array}{ccc|ccc} -1 & 2 & 2 & 1 & 0 & 0 \\ -1 & 2 & 5 & 0 & 1 & 0 \\ -2 & 2 & 10 & 0 & 0 & 1 \end{array}\right] \xrightarrow[-2①+③]{-①+②} \left[\begin{array}{ccc|ccc} -1 & 2 & 2 & 1 & 0 & 0 \\ 0 & 0 & 3 & -1 & 1 & 0 \\ 0 & -2 & 6 & -2 & 0 & 1 \end{array}\right] \xrightarrow[②\leftrightarrow③]{-①}$$

$$\left[\begin{array}{ccc|ccc} 1 & -2 & -2 & -1 & 0 & 0 \\ 0 & -2 & 6 & -2 & 0 & 1 \\ 0 & 0 & 3 & -1 & 1 & 0 \end{array}\right] \xrightarrow[\frac{1}{3}③]{-\frac{1}{2}②} \left[\begin{array}{ccc|ccc} 1 & -2 & -2 & -1 & 0 & 0 \\ 0 & 1 & -3 & 1 & 0 & -\frac{1}{2} \\ 0 & 0 & 1 & -\frac{1}{3} & \frac{1}{3} & 0 \end{array}\right] \xrightarrow[2③+①]{3③+②}$$

$$\left[\begin{array}{ccc|ccc} 1 & -2 & 0 & -\frac{5}{3} & \frac{2}{3} & 0 \\ 0 & 1 & 0 & 0 & 1 & -\frac{1}{2} \\ 0 & 0 & 1 & -\frac{1}{3} & \frac{1}{3} & 0 \end{array}\right] \xrightarrow{2②+①} \left[\begin{array}{ccc|ccc} 1 & 0 & 0 & -\frac{5}{3} & \frac{8}{3} & -1 \\ 0 & 1 & 0 & 0 & 1 & -\frac{1}{2} \\ 0 & 0 & 1 & -\frac{1}{3} & \frac{1}{3} & 0 \end{array}\right]$$

17. $\begin{bmatrix} 2 & 1 \\ 1 & -1 \end{bmatrix} \begin{bmatrix} x \\ y \end{bmatrix} = \begin{bmatrix} 4 \\ 0 \end{bmatrix}$

$\left[\begin{array}{cc|cc} 2 & 1 & 1 & 0 \\ 1 & -1 & 0 & 1 \end{array}\right] \xrightarrow{-②+①} \left[\begin{array}{cc|cc} 1 & 2 & 1 & 1 \\ 1 & -1 & 0 & 1 \end{array}\right] \xrightarrow{-①+②}$

$\left[\begin{array}{cc|cc} 1 & 2 & 1 & 1 \\ 0 & -3 & -1 & 0 \end{array}\right] \xrightarrow{-\frac{1}{3}②} \left[\begin{array}{cc|cc} 1 & 2 & 1 & 1 \\ 0 & 1 & \frac{1}{3} & 0 \end{array}\right] \xrightarrow{-2②+①}$

$\left[\begin{array}{cc|cc} 1 & 0 & \frac{1}{3} & 1 \\ 0 & 1 & \frac{1}{3} & 0 \end{array}\right]$

$\begin{bmatrix} \frac{1}{3} & 1 \\ \frac{1}{3} & 0 \end{bmatrix} \begin{bmatrix} 4 \\ 0 \end{bmatrix} = \begin{bmatrix} \frac{4}{3} \\ \frac{4}{3} \end{bmatrix}$

19. $\begin{bmatrix} 2 & 0 & 1 \\ 0 & 3 & 1 \\ -2 & 0 & 4 \end{bmatrix} \begin{bmatrix} x \\ y \\ z \end{bmatrix} = \begin{bmatrix} 1 \\ 2 \\ 0 \end{bmatrix}$

$\left[\begin{array}{ccc|ccc} 2 & 0 & 1 & 1 & 0 & 0 \\ 0 & 3 & 1 & 0 & 1 & 0 \\ -2 & 0 & 4 & 0 & 0 & 1 \end{array}\right] \begin{array}{c} \xrightarrow{①+③} \\ \frac{1}{3}② \end{array} \left[\begin{array}{ccc|ccc} 2 & 0 & 1 & 1 & 0 & 0 \\ 0 & 1 & \frac{1}{3} & 0 & \frac{1}{3} & 0 \\ 0 & 0 & 5 & 1 & 0 & 1 \end{array}\right] \begin{array}{c} \xrightarrow{\frac{1}{2}①} \\ \frac{1}{5}③ \end{array}$

$\left[\begin{array}{ccc|ccc} 1 & 0 & \frac{1}{2} & \frac{1}{2} & 0 & 0 \\ 0 & 1 & \frac{1}{3} & 0 & \frac{1}{3} & 0 \\ 0 & 0 & 1 & \frac{1}{5} & 0 & \frac{1}{5} \end{array}\right] \begin{array}{c} \xrightarrow{-\frac{1}{2}③+①} \\ -\frac{1}{3}③+② \end{array} \left[\begin{array}{ccc|ccc} 1 & 0 & 0 & \frac{2}{5} & 0 & -\frac{1}{10} \\ 0 & 1 & 0 & -\frac{1}{15} & \frac{1}{3} & -\frac{1}{15} \\ 0 & 0 & 1 & \frac{1}{5} & 0 & \frac{1}{5} \end{array}\right]$

$\begin{bmatrix} \frac{2}{5} & 0 & -\frac{1}{10} \\ -\frac{1}{15} & \frac{1}{3} & -\frac{1}{15} \\ \frac{1}{5} & 0 & \frac{1}{5} \end{bmatrix} \begin{bmatrix} 1 \\ 2 \\ 0 \end{bmatrix} = \begin{bmatrix} \frac{2}{5} \\ \frac{3}{5} \\ \frac{1}{5} \end{bmatrix}$

21.
$$\begin{bmatrix} 1 & 0 & 1 \\ 2 & 1 & 2 \\ -1 & 0 & 0 \end{bmatrix} \begin{bmatrix} x \\ y \\ z \end{bmatrix} = \begin{bmatrix} - \\ 2 \\ 1 \end{bmatrix}$$

$$A^{-1} = \begin{bmatrix} 0 & 0 & -1 \\ -2 & 1 & 0 \\ 1 & 0 & 1 \end{bmatrix} \qquad \text{(See 6.5, \#7)}$$

$$\begin{bmatrix} 0 & 0 & -1 \\ -2 & 1 & 0 \\ 1 & 0 & 1 \end{bmatrix} \begin{bmatrix} 0 \\ 2 \\ 1 \end{bmatrix} = \begin{bmatrix} -1 \\ 2 \\ 1 \end{bmatrix}$$

23.
$$\begin{bmatrix} -1 & 2 & 2 \\ -1 & 2 & 5 \\ -1 & 1 & 5 \end{bmatrix} \begin{bmatrix} x \\ y \\ z \end{bmatrix} = \begin{bmatrix} 0 \\ 0 \\ -2 \end{bmatrix}$$

Multiply row 3 by 2,

$$\begin{bmatrix} -1 & 2 & 2 \\ -1 & 2 & 5 \\ -2 & 2 & 10 \end{bmatrix} \begin{bmatrix} x \\ y \\ z \end{bmatrix} = \begin{bmatrix} 0 \\ 0 \\ -4 \end{bmatrix}$$

$$A^{-1} = \begin{bmatrix} -\frac{5}{3} & \frac{8}{3} & -1 \\ 0 & 1 & -\frac{1}{2} \\ -\frac{1}{3} & \frac{1}{3} & 0 \end{bmatrix} \qquad \text{(See 6.5, \#15)}$$

$$\begin{bmatrix} -\frac{5}{3} & \frac{8}{3} & -1 \\ 0 & 1 & -\frac{1}{2} \\ -\frac{1}{3} & \frac{1}{3} & 0 \end{bmatrix} \begin{bmatrix} 0 \\ 0 \\ -4 \end{bmatrix} = \begin{bmatrix} 4 \\ 2 \\ 0 \end{bmatrix}$$

25.
$$\begin{bmatrix} 1 & 1 & 5 \\ 3 & 8 & 5 \\ -1 & -5 & 3 \end{bmatrix} \begin{bmatrix} x \\ y \\ z \end{bmatrix} = \begin{bmatrix} 2 \\ 0 \\ 15 \end{bmatrix}$$

$$\left[\begin{array}{ccc|ccc} 1 & 1 & 5 & 1 & 0 & 0 \\ 3 & 8 & 5 & 0 & 1 & 0 \\ -1 & -5 & 3 & 0 & 0 & 1 \end{array}\right] \begin{array}{c} -3①+② \\ \hline ①+③ \end{array} \left[\begin{array}{ccc|ccc} 1 & 1 & 5 & 1 & 0 & 0 \\ 0 & 5 & 10 & -3 & 1 & 0 \\ 0 & -4 & 8 & 1 & 0 & 1 \end{array}\right] \frac{1}{5}② \longrightarrow$$

$$\left[\begin{array}{ccc|ccc} 1 & 1 & 5 & 1 & 0 & 0 \\ 0 & 1 & 2 & -\frac{3}{5} & \frac{1}{5} & 0 \\ 0 & -4 & 8 & 1 & 0 & 1 \end{array}\right] \begin{array}{c} 4②+③ \\ \hline \end{array} \left[\begin{array}{ccc|ccc} 1 & 1 & 5 & 1 & 0 & 0 \\ 0 & 1 & -2 & -\frac{3}{5} & \frac{1}{5} & 0 \\ 0 & 0 & 0 & -\frac{7}{5} & -\frac{4}{5} & 1 \end{array}\right]$$

No solution

Section 6.6 Determinants

1. $\begin{vmatrix} 2 & 4 \\ 3 & 5 \end{vmatrix} = 10 - 12 = -2$

3. $\begin{vmatrix} 0 & 1 \\ 2 & 6 \end{vmatrix} = 0 - 2 = -2$

5. $\begin{vmatrix} 2 & -6 \\ 5 & -3 \end{vmatrix} = -6 + 30 = 24$

7. $\begin{vmatrix} 5 & 10 \\ 2 & 4 \end{vmatrix} = 20 - 20 = 0$

9. $\begin{vmatrix} 2 & 0 & 0 \\ 3 & 2 & 4 \\ 1 & -3 & 5 \end{vmatrix} = 20 + 0 + 0 - (0-24+0) = 44$

$5(4-0) \quad 2(10-0)$

$2 \begin{vmatrix} 2 & 4 \\ -3 & 5 \end{vmatrix} = 2(10+12) = 44$

11.

$$\begin{vmatrix} 2 & 3 & 4 \\ 2 & 0 & 3 \\ -1 & 6 & 5 \end{vmatrix} = 0 - 9 - 48 - (0+36+30) = -123$$

By row 2:

$$-2\begin{vmatrix} 3 & -4 \\ 6 & 5 \end{vmatrix} -3\begin{vmatrix} 2 & 3 \\ -1 & 6 \end{vmatrix} = -2(15+24) - 3(12+3) = -123$$

13.

$$\begin{vmatrix} 3 & 1 & 9 \\ 6 & 1 & -2 \\ -4 & 1 & 5 \end{vmatrix} = 15 + 8 + 54 - (-36-6+30) = 89$$

By column 2:

$$-1\begin{vmatrix} 6 & -2 \\ -4 & 5 \end{vmatrix} +1\begin{vmatrix} 3 & 9 \\ -4 & 5 \end{vmatrix} -1\begin{vmatrix} 3 & 9 \\ 6 & -2 \end{vmatrix}$$

$$= -(30-8) + (15+36) - (-6-54) = 89$$

15.

$$\begin{vmatrix} 6 & 7 & 6 \\ 2 & 5 & -2 \\ 3 & 1 & -3 \end{vmatrix} = -90 + 42 + 12 - (90+12-42) = -96$$

By row 3:

$$3\begin{vmatrix} 7 & 6 \\ 5 & 2 \end{vmatrix} -1\begin{vmatrix} 6 & 6 \\ 2 & 2 \end{vmatrix} -3\begin{vmatrix} 6 & 7 \\ 2 & 5 \end{vmatrix}$$

$$= 3(14-30) -1(12-12) -3(30-14) = -96$$

17.

$$\begin{vmatrix} 9 & -6 & 0 \\ -1 & -3 & -1 \\ -5 & -4 & -7 \end{vmatrix} = 189 - 30 + 0 - (0+36-42) = 165$$

By row 1:

$$9\begin{vmatrix} -3 & -1 \\ -4 & -7 \end{vmatrix} +6\begin{vmatrix} -1 & -1 \\ -5 & -7 \end{vmatrix} = 9(21-4) +6(7-5) = 165$$

19.
$$\begin{vmatrix} x & 1 \\ 1 & x \end{vmatrix} = 0$$

$$x^2 - 1 = 0$$

$$x = \pm 1$$

21.
$$\begin{Vmatrix} \begin{matrix} 3 & 1 \\ -1 & 2 \\ 5 & 0 \\ 0 & 1 \end{matrix} & \begin{matrix} 7 & 0 \\ 1 & 2 \\ 0 & 0 \\ 0 & 1 \end{matrix} \end{Vmatrix} = \begin{vmatrix} 7 & 14 \\ 5 & 0 \end{vmatrix} = -70$$

23.
$$\begin{vmatrix} x & 2 & 1 \\ -1 & 3 & 0 \\ 1 & x & 0 \end{vmatrix} = 0$$

By column 3:

$$0 = 1 \begin{vmatrix} -1 & 3 \\ 1 & x \end{vmatrix} = -x - 3$$

$$x = -3$$

25.
$$\begin{vmatrix} 1 & 4 & -2 & 0 \\ 0 & 1 & 1 & 2 \\ -3 & -2 & 1 & 0 \\ -1 & 1 & 2 & 3 \end{vmatrix} = 2 \begin{vmatrix} 1 & 4 & -2 \\ -3 & -2 & 1 \\ -1 & 1 & 2 \end{vmatrix} + 3 \begin{vmatrix} 1 & 4 & -2 \\ 0 & 1 & 1 \\ -3 & -2 & 1 \end{vmatrix}$$

$$= \left[-4-4+6 - (-4+1-24) \right] + 3 \left[1+0-12 - (6-2+0) \right] = 5$$

27a. If a row (col) of A has all zeros, then $\det A = 0$.

$$\det A = \begin{vmatrix} a_{11} & 0 \\ a_{21} & 0 \end{vmatrix} = a_{11}(0) - a_{21}(0) = 0$$

27b. If A is obtained by interchanging two rows (cols), then $\det A = -\det B$.

$$\det B = \begin{vmatrix} a_{21} & a_{22} \\ a_{11} & a_{12} \end{vmatrix} = a_{21}a_{12} - a_{11}a_{22} = -(a_{11}a_{22} - a_{21}a_{12}) = -\begin{vmatrix} a_{11} & a_{12} \\ a_{21} & a_{22} \end{vmatrix}$$

$$= -\det A$$

27c. If any two rows (cols) of A are equal, then $\det A = 0$.

$$\begin{vmatrix} a_{11} & a_{12} \\ a_{11} & a_{12} \end{vmatrix} = a_{11}a_{12} - a_{11}a_{12} = 0$$

27d. If A and B are identical except that a row (col) of B is k times the same row (col) of A, then $\det A = k \det B$.

$$\det B = \begin{vmatrix} ka_{11} & ka_{12} \\ a_{21} & a_{22} \end{vmatrix} = ka_{11}a_{22} - ka_{21}a_{12} = k(a_{11}a_{22} - a_{21}a_{12}) = k \det A$$

27e. The value of the determinant is unchanged if the rows and columns of the matrix are interchanged.

$$\begin{vmatrix} a_{11} & a_{12} \\ a_{21} & a_{22} \end{vmatrix} = a_{11}a_{22} - a_{12}a_{21} = \begin{vmatrix} a_{11} & a_{21} \\ a_{12} & a_{22} \end{vmatrix}$$

27f. The value of a determinant is unchanged if a multiple of one row (col) is added to another row (col.)

$$\det B = \begin{vmatrix} a_{11} & a_{12} \\ ka_{11}+a_{21} & ka_{12}+a_{22} \end{vmatrix} = a_{11}(ka_{12}+a_{22}) - a_{12}(ka_{11}+a_{21})$$

$$= a_{11}a_{22} - a_{12}a_{21} = \det A$$

1. $x = \dfrac{\begin{vmatrix} 2 & -4 \\ -4 & 1 \end{vmatrix}}{\begin{vmatrix} 1 & -4 \\ -2 & 1 \end{vmatrix}} = \dfrac{-14}{-9} = \dfrac{14}{9}$ $y = \dfrac{\begin{vmatrix} 1 & 2 \\ -2 & -4 \end{vmatrix}}{-9} = 0$

$1 - (+8)$

3. $x = \dfrac{\begin{vmatrix} 23 & 4 \\ -1 & -3 \end{vmatrix}}{\begin{vmatrix} 3 & 4 \\ 1 & -3 \end{vmatrix}} = \dfrac{-65}{-13} = 5,$ $y = \dfrac{\begin{vmatrix} 3 & 23 \\ 1 & -1 \end{vmatrix}}{-13} = \dfrac{-26}{-13} = 2$

$-9 \cdot \left(4\right)$

5. $x = \dfrac{\begin{vmatrix} 8 & -4 \\ 16 & -8 \end{vmatrix}}{\begin{vmatrix} 1 & -4 \\ 2 & -8 \end{vmatrix}} = \dfrac{0}{0},$ Dependent

7. $x = \dfrac{\begin{vmatrix} 10 & 4 \\ -2 & 8 \end{vmatrix}}{\begin{vmatrix} 3 & 4 \\ 6 & 8 \end{vmatrix}} = \dfrac{88}{0},$ Inconsistent

9. $x = \dfrac{\begin{vmatrix} 2 & -1 \\ 6 & 1 \end{vmatrix}}{\begin{vmatrix} 3 & -1 \\ 2 & 1 \end{vmatrix}} = \dfrac{8}{5},$ $y = \dfrac{\begin{vmatrix} 3 & 2 \\ 2 & 6 \end{vmatrix}}{5} = \dfrac{14}{5}$

11. $x = \dfrac{\begin{vmatrix} 2 & -1 \\ -5 & 1 \end{vmatrix}}{\begin{vmatrix} 2 & -1 \\ 1 & 1 \end{vmatrix}} = \dfrac{-3}{3} = -1,$ $\qquad y = \dfrac{\begin{vmatrix} 2 & 2 \\ 1 & -5 \end{vmatrix}}{3} = \dfrac{-12}{3} = -4$

13. $x = \dfrac{\begin{vmatrix} 8 & 1 & 1 \\ 12 & 1 & -1 \\ 2 & -1 & 1 \end{vmatrix}}{\begin{vmatrix} 1 & 1 & 1 \\ 1 & 1 & -1 \\ 1 & -1 & 1 \end{vmatrix}} = \dfrac{-28}{-4} = 7,$ $\qquad y = \dfrac{\begin{vmatrix} 1 & 8 & 1 \\ 1 & 12 & -1 \\ 1 & 2 & 1 \end{vmatrix}}{-4} = \dfrac{-12}{-4} = 3,$

$z = \dfrac{\begin{vmatrix} 1 & 1 & 8 \\ 1 & 1 & 12 \\ 1 & -1 & 2 \end{vmatrix}}{-4} = \dfrac{8}{-4} = -2$

15. $r = \dfrac{\begin{vmatrix} 2 & 3 & 2 \\ -1 & 2 & -1 \\ 0 & -3 & 3 \end{vmatrix}}{\begin{vmatrix} 1 & 3 & 2 \\ 3 & 2 & -1 \\ 1 & -3 & -3 \end{vmatrix}} = \dfrac{21}{-7} = -3,$ $\qquad s = \dfrac{\begin{vmatrix} 1 & 2 & 2 \\ 3 & -1 & -1 \\ 1 & 0 & 3 \end{vmatrix}}{-7} = \dfrac{-21}{-7} = 3$

$t = \dfrac{\begin{vmatrix} 1 & 3 & 2 \\ 3 & 2 & -1 \\ 1 & -3 & 0 \end{vmatrix}}{-7} = \dfrac{-28}{-7} = 4$

17. $x = \dfrac{\begin{vmatrix} 5 & -3 & 4 \\ 4 & -7 & 4 \\ 2 & 3 & -5 \end{vmatrix}}{\begin{vmatrix} 2 & -3 & 4 \\ 5 & -7 & 4 \\ -2 & 3 & -5 \end{vmatrix}} = \dfrac{135}{-1} = -135, \qquad y = \dfrac{\begin{vmatrix} 2 & 5 & 4 \\ 5 & 4 & 4 \\ -2 & 2 & -5 \end{vmatrix}}{-1} = \dfrac{101}{1} = -101$

$$z = \dfrac{\begin{vmatrix} 2 & -3 & 5 \\ 5 & -7 & 4 \\ -2 & 3 & 2 \end{vmatrix}}{-1} = \dfrac{7}{-1} = -7$$

19. $x = \dfrac{\begin{vmatrix} 0 & 2 & -1 \\ 0 & -1 & 1 \\ 0 & -3 & -2 \end{vmatrix}}{\begin{vmatrix} 1 & 2 & -1 \\ 1 & -1 & 1 \\ 2 & -3 & -2 \end{vmatrix}} = \dfrac{0}{14} = 0, \quad y = 0, \quad z = 0$

21. $u = \dfrac{\begin{vmatrix} 8 & -4 & 2 \\ 1 & 2 & 1 \\ 17 & 0 & -18 \end{vmatrix}}{\begin{vmatrix} 9 & -4 & 2 \\ 3 & 2 & 1 \\ 12 & 0 & -18 \end{vmatrix}} = \dfrac{-496}{-636}, \qquad v = \dfrac{\begin{vmatrix} 9 & 8 & 2 \\ 3 & 1 & 1 \\ 12 & 17 & -18 \end{vmatrix}}{-636} = \dfrac{291}{-636}$

$$w = \dfrac{\begin{vmatrix} 9 & -4 & 8 \\ 3 & 2 & 1 \\ 12 & 0 & 17 \end{vmatrix}}{-636} = \dfrac{270}{-636}$$

23. $w + 6 = l, \quad w - l = 6$

$2w + 2l = 40$

$$w = \frac{\begin{vmatrix} 6 & -1 \\ 40 & 2 \end{vmatrix}}{\begin{vmatrix} 1 & -1 \\ 2 & 2 \end{vmatrix}} = \frac{52}{4} = 13 \qquad l = \frac{\begin{vmatrix} 1 & 6 \\ 2 & 40 \end{vmatrix}}{4} = 7$$

25.

	Rate	Time
R_1	50	$t+1$
R_2	60	t

$50(t+1) = d, \quad 50t - d = -50$

$60t \quad\quad = d, \quad 60t - d = \quad 0$

$$t = \frac{\begin{vmatrix} -50 & -1 \\ 0 & -1 \end{vmatrix}}{\begin{vmatrix} 50 & -1 \\ 60 & -1 \end{vmatrix}} = \frac{50}{10} = 5$$

27. $5n + 10d + 25q = 1900$

$\quad n + \quad d + \quad q = 100$

$\quad n - \quad 3d \quad\quad = \quad 0$

$$d = \frac{\begin{vmatrix} 5 & 1900 & 25 \\ 1 & 100 & 1 \\ 1 & 0 & 0 \end{vmatrix}}{\begin{vmatrix} 5 & 10 & 25 \\ 1 & 1 & 1 \\ 1 & -3 & 0 \end{vmatrix}} = \frac{-600}{-65} = 8$$

29.

$$x + y + z = 50 \qquad x + y + z = 50$$
$$x = y + 5 \qquad x - y = 5$$
$$x = z + 10 \qquad x - z = 10$$

$$x = \frac{\begin{vmatrix} 50 & 1 & 1 \\ 5 & -1 & 0 \\ 10 & 0 & -1 \end{vmatrix}}{\begin{vmatrix} 1 & 1 & 1 \\ 1 & -1 & 0 \\ 1 & 0 & -1 \end{vmatrix}} = \frac{65}{3}, \qquad y = \frac{\begin{vmatrix} 1 & 50 & 1 \\ 1 & 5 & 0 \\ 1 & 10 & -1 \end{vmatrix}}{3} = \frac{50}{3}$$

$$z = \frac{\begin{vmatrix} 1 & 1 & 50 \\ 1 & -1 & 5 \\ 1 & 0 & 10 \end{vmatrix}}{3} = \frac{35}{3}$$

31.

$$u' = \frac{\begin{vmatrix} F & Q_1 \\ 0 & Q_2 \end{vmatrix}}{\begin{vmatrix} R_1 & Q_1 \\ R_2 & Q_2 \end{vmatrix}} = \frac{FQ_2}{R_1 Q_2 - R_2 Q_1}$$

$$v' = \frac{\begin{vmatrix} R_1 & F \\ R_2 & 0 \end{vmatrix}}{\begin{vmatrix} R_1 & Q_1 \\ R_2 & Q_2 \end{vmatrix}} = \frac{-R_2 F}{R_1 Q_2 - R_2 Q_1}$$

33. $(1, 0) \rightarrow 0 = 1 + a + b + c$ $\qquad$ $a + b = -2$

$\qquad$ $(3, 0) \rightarrow 0 = 27 + 9a + 3b + c$ $\qquad$ $9a + 3b = -28$

$\qquad$ $(0, 1) \rightarrow 1 = c$

$$a = \dfrac{\begin{vmatrix} -2 & 1 \\ -28 & 3 \end{vmatrix}}{\begin{vmatrix} 1 & 1 \\ 9 & 3 \end{vmatrix}} = \dfrac{-22}{-6} = \dfrac{11}{3}, \qquad b = \dfrac{\begin{vmatrix} 1 & -2 \\ 9 & -28 \end{vmatrix}}{-6} = \dfrac{-10}{-6} = \dfrac{5}{3}$$

35. $(1, 2) \rightarrow 2 = \dfrac{a}{1+c}$, $\qquad$ $(2, 1) \rightarrow 1 = \dfrac{2a}{2+c}$

$\qquad$ $a - 2c = 2$

$\qquad$ $2a - c = 2$

$$a = \dfrac{\begin{vmatrix} 2 & -2 \\ 2 & -1 \end{vmatrix}}{\begin{vmatrix} 1 & -2 \\ 2 & -1 \end{vmatrix}} = \dfrac{2}{3}, \qquad c = \dfrac{\begin{vmatrix} 1 & 2 \\ 2 & 2 \end{vmatrix}}{3} = \dfrac{-2}{3}$$

Section 6.8 Linear Inequalities in Two Unknowns

5. $x + y > x - y$

$\qquad$ $2y > 0$

$\qquad$ $y > 0$

7. $2x - y - 8 > x - 2y + 8$

$\qquad$ $x + y > 16$

9. $3(x-1) + 4x \geq 7x + y + 2(x-y)$

$\qquad$ $3x - 3 + 4x \geq 7x + y + 2x - 2y$

11. $|y| < 3$

$\qquad$ $y < 3$ $\qquad$ or $\qquad$ $y > -3$

13. $|x| < y$

$\qquad$ $x < y$ $\qquad$ or $\qquad$ $x > -y$

15. $|x| < 1-|y|$

    a. $x \geq 0$ and $y \geq 0$ **b.** $x \geq 0$ and $y < 0$

            $x < 1-y$ $x < 1+y$

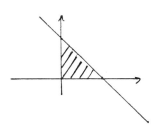
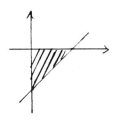

    c. $x < 0$ and $y \geq 0$ **d.** $x < 0$ and $y < 0$

           $-x < 1-y$ $-x < 1+y$

             $x > y-1$ $x > -1 - y$

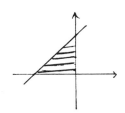
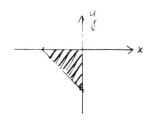

17. $\frac{9}{5}C+32 < 0$ 19. $35x + 15y \leq 1000$

       $\frac{9}{5}C < -32$ $x = 0,$ $y \leq \frac{1000}{15} \simeq 67$

        $C < -\frac{160}{9}$ $y = 0,$ $x \leq \frac{1000}{35} \simeq 29$

Sections 6.9 and 6.10   Systems of Linear Inequalities

                         Linear Programming

17. $y = 2$

   $x + 2y = 8$

   $x = 8-4 = 4, (4.2)$

   $p = x+3y$

   $(0, 2)$ $p = 0 + 3(2) = 6$

   $(0, 0)$ $p = 0 + 0 = 0$

   $(4, 2)$ $p = 4 + 3(2) = 10$, maximum

   $(8, 0)$ $p = 8 + 0 = 8$

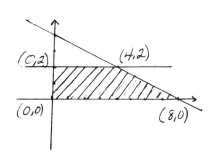

19. $x + y = 5$
$x + 2y = 8$

$- y = -3, \quad y = 3, \quad x = 2$

$p = 5x + y$

$(0, 4) \quad p = 5(0) + 4 = 4$

$(0, 0) \quad p = 5(0) + 0 = 0$

$(2, 3) \quad p = 5(2) + 3 = 13$

$(5, 0) \quad p = 5(5) + 0 = 25$, maximum

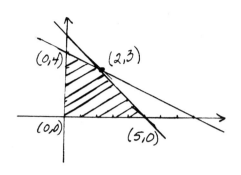

21. $x + 3y = 8$
$2x + y = 6$

$2x + y = 6$
$-2x - 6y = -16$

$- 5y = -10,$

$y = 2, \quad x = 2$

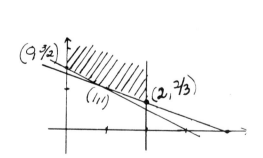

a. $p = 2x + 3y$

$(0, 0) \quad p = 0 + 0 = 0$

$(0, \frac{8}{3}) \quad p = 0 + 3(\frac{8}{3}) = 8$

$(3, 0) \quad p = 2(3) + 0 = 6$

$(2, 2) \quad p = 2(2) + 3(2) = 10$, maximum

b. $p = 3x + y$

$(0, 0) \quad p = 0 + 0 = 0$

$(0, \frac{8}{3}) \quad p = 3(0) + (\frac{8}{3}) = \frac{8}{3}$

$(3, 0) \quad p = 3(3) + 9 = 9$, maximum

$(2, 2) \quad p = 3(2) + 2 = 8$

23. $x + 2y = 3$
$x + 3y = 4$

$y = 1, \quad x = 1$

$p = 5x + 3y$

$(1, 1) \quad p = 5(1) + 3(1) = 8$

$(0, \frac{3}{2}) \quad p = 5(0) + 3(\frac{3}{2}) = \frac{9}{2}$, minimum

$(2, \frac{2}{3}) \quad p = 5(2) + 3(\frac{3}{2}) = 12$

25. $2x + y = 5$
$x + 2y = 4$

$2x + y = 5$
$-2x - 4y = -8$

$y = 1, \quad x = 2$

$p = 3x + y$

$(0, 2) \quad p = 3(0) + 2 = 2$

$(2, 1) \quad p = 3(2) + 1 = 7$

$(0, 0) \quad p = 0$

$(\frac{5}{2}, 0) \quad p = 3(\frac{5}{2}) + 0 = \frac{15}{2}$, maximum

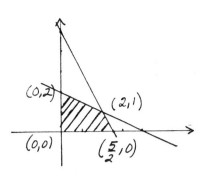

27. $2x + 3y = 120$
$4x + y = 90$

$2x + 3y = 120$
$-12x - 3y = -270$

$-10x = -150$

$x = 15, \quad y = 30$

$p = 10x + 8y$

$(0, 40) \quad p = 10(0) + 8(40) = 320$

$(15, 30) \quad p = 10(15) + 8(30) = 390$

$(\frac{45}{2}, 0) \quad p = 10(\frac{45}{2}) + 0 = 225$

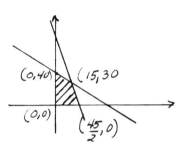

29. $x \geq 0, \quad y \geq 0, \quad x + y \leq 20,000$

$y \leq 2x,$

$I = .07x + .085y$

$x + y = 20,000$
$2x - y = 0$

$3x = 20,000, \qquad x = 6700$

$y = 13,300$

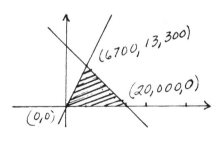

$(0, 0) \qquad I = 0$

$(20,000, 0) \qquad I = .07(20,000) + 0 = 1400$

$(6700, 13,300) \qquad I = .07(6700) + .085(13,300) = 1179.5$

1. $x + 3y = 5$
$$\underline{2x - 3y = 1}$$

$3x = 6, \quad x = 2$
$3y = 5 - x = 3, \quad y = 1$

3. $2s + 5t = -4$
$$\underline{6s + 8t = 10}$$

$-6s - 15t = 12$
$$\underline{6s + 8t = 10}$$

$-7t = 22, \quad t = -\frac{22}{7}$

$2s = -4 - 5\left(-\frac{22}{7}\right), \quad s = \frac{41}{7}$

5. $5x + 6z + 3 = 0$
$$\underline{7x + 4z - 9 = 0}$$

$10x + 12z + 6 = 0$
$$\underline{-21x - 12z + 27 = 0}$$

$-11x = -33, \quad x = 3$

$6z = -5x - 3, \quad z = -3$

7. $x + y + z = 1$

$x + y - z = 2$

$$\underline{x - y + z = 3}$$

$x + y + z = 1$
$$\underline{x + y - z = 2}$$

$2x + 2y = 3$

$x + y - z = 2$
$$\underline{x - y + z = 3}$$

$2x = 5, \quad x = \frac{5}{2}$

$2y = 3 - 2x = 3 - 2\left(\frac{5}{2}\right),$

$y = -1$

$z = 1 - 1 - y = 1 - \frac{5}{2} + 1$

$= -\frac{1}{2}$

9. $a + 2b - 3c = 1$

$b - c = 4$

$$\underline{-a + b + c = 5}$$

$a + 2b - 3c = 1$
$$\underline{-a + b + c = 5}$$

$3b - 2c = 6$
$$\underline{2b - 2c = 8}$$

$b = -2$

$c = b - 4 = -2 - 4 = -6$

$a = b + c - 5 = -2 - 6 - 5 = -13$

11. $\begin{bmatrix} 3 & 5 & | & 13 \\ -1 & 3 & | & 3 \end{bmatrix} \xrightarrow{\frac{1}{3}①+②} \begin{bmatrix} 3 & 5 & | & 13 \\ 0 & \frac{14}{3} & | & \frac{28}{3} \end{bmatrix} \xrightarrow{\frac{3}{14}②} \begin{bmatrix} 3 & 5 & | & 13 \\ 0 & 1 & | & 2 \end{bmatrix} \xrightarrow{-5②+①}$

$\begin{bmatrix} 3 & 0 & | & 3 \\ 0 & 1 & | & 2 \end{bmatrix} \xrightarrow{\frac{1}{3}①} \begin{bmatrix} 1 & 0 & | & 1 \\ 0 & 1 & | & 2 \end{bmatrix}$

13.
$$\begin{bmatrix} 1 & -2 \\ 1 & 2 \end{bmatrix} \begin{bmatrix} x \\ y \end{bmatrix} = \begin{bmatrix} 1 \\ 3 \end{bmatrix}$$

$$\left[\begin{array}{cc|cc} 1 & -2 & 1 & 0 \\ 1 & 2 & 0 & 1 \end{array}\right] \xrightarrow{-①+②} \left[\begin{array}{cc|cc} 1 & -2 & 1 & 0 \\ 0 & 4 & -1 & 1 \end{array}\right] \xrightarrow{\frac{1}{2}②+①}$$

$$\left[\begin{array}{cc|cc} 1 & 0 & \frac{1}{2} & \frac{1}{2} \\ 0 & 4 & -1 & 1 \end{array}\right] \xrightarrow{\frac{1}{4}②} \left[\begin{array}{cc|cc} 1 & 0 & \frac{1}{2} & \frac{1}{2} \\ 0 & 1 & -\frac{1}{4} & \frac{1}{4} \end{array}\right]$$

$$\begin{bmatrix} \frac{1}{2} & \frac{1}{2} \\ -\frac{1}{4} & \frac{1}{4} \end{bmatrix} \begin{bmatrix} 1 \\ 3 \end{bmatrix} = \begin{bmatrix} 2 \\ \frac{1}{2} \end{bmatrix}$$

15.
$$\left[\begin{array}{ccc|c} 1 & -2 & 3 & 4 \\ -3 & 4 & -1 & -2 \\ 2 & 1 & -4 & 3 \end{array}\right] \xrightarrow{②+③} \left[\begin{array}{ccc|c} 1 & -2 & 3 & 4 \\ -3 & 4 & -1 & -2 \\ -1 & 5 & -5 & 1 \end{array}\right] \xrightarrow{①+③}$$

$$\left[\begin{array}{ccc|c} 1 & -2 & 3 & 4 \\ -3 & 4 & -1 & -2 \\ 0 & 3 & -2 & 5 \end{array}\right] \xrightarrow{3①+②} \left[\begin{array}{ccc|c} 1 & -2 & 3 & 4 \\ 0 & -2 & 8 & 10 \\ 0 & 3 & -2 & 5 \end{array}\right] \xrightarrow[②+③]{-②+①}$$

$$\left[\begin{array}{ccc|c} 1 & 0 & -5 & -6 \\ 0 & -2 & 8 & 10 \\ 0 & 1 & 6 & 15 \end{array}\right] \xrightarrow{2③+②} \left[\begin{array}{ccc|c} 1 & 0 & -5 & -6 \\ 0 & 0 & 20 & 40 \\ 0 & 1 & 6 & 15 \end{array}\right] \xrightarrow{\frac{1}{20}②}$$

$$\left[\begin{array}{ccc|c} 1 & 0 & -5 & -6 \\ 0 & 0 & 1 & 2 \\ 0 & 1 & 6 & 15 \end{array}\right] \xrightarrow[-6②+③]{5②+①} \left[\begin{array}{ccc|c} 1 & 0 & 0 & 4 \\ 0 & 0 & 1 & 2 \\ 0 & 1 & 0 & 3 \end{array}\right]$$

– 105 –

17. $y + x = -6$

$y + x^2 = 0, \quad y = -x^2$

$-x^2 + x = -6$

$x^2 - x - 6 = 0$

$(x-3)(x+2) = 0$

$x = -2, 3$

$y = -4, -9$

$(-2, 4), \quad (3, -9)$

19. $x^2 + y^2 = 6$

$\underline{x^2 - y^2 = 2}$

$2x^2 = 8$

$x^2 = 4, \quad x = \pm 2$

$y^2 = 6 - x^2$

$y^2 = 2$

$y = \pm\sqrt{2}$

$(\pm 2, \pm\sqrt{2})$

23.
$$\begin{bmatrix} 1 & 1 & 2 \\ 2 & 3 & 2 \\ 0 & 4 & -1 \end{bmatrix} + \begin{bmatrix} -3 & 7 & 6 \\ 0 & -2 & 3 \\ -2 & 7 & 1 \end{bmatrix} = \begin{bmatrix} -2 & 8 & 8 \\ 2 & 1 & 5 \\ -2 & 11 & 0 \end{bmatrix}$$

25. $AB = \begin{bmatrix} 2 & 3 & -1 \end{bmatrix} \begin{bmatrix} -1 \\ 2 \\ 4 \end{bmatrix} = \begin{bmatrix} 2(-1)+3(2)-1(4) \end{bmatrix} = \begin{bmatrix} 0 \end{bmatrix}$

$BA = \begin{bmatrix} -1 \\ 2 \\ 4 \end{bmatrix} \begin{bmatrix} 2 & 3 & -1 \end{bmatrix} = \begin{bmatrix} -1(2) & -1(3) & -1(-1) \\ 2(2) & 2(3) & 2(-1) \\ 4(2) & 4(3) & 4(-1) \end{bmatrix}$

$= \begin{bmatrix} -2 & -3 & 1 \\ 4 & 6 & -2 \\ 8 & 12 & -4 \end{bmatrix}$

27. $AB = \begin{bmatrix} 2 & -1 \\ 3 & 0 \end{bmatrix} \begin{bmatrix} 5 & -3 \\ 1 & 7 \end{bmatrix} = \begin{bmatrix} 2(5)-1(1) & 2(-3)-1(7) \\ 3(5)+0(1) & 3(-3)+0(7) \end{bmatrix} = \begin{bmatrix} 9 & -13 \\ 15 & -9 \end{bmatrix}$

$BA = \begin{bmatrix} 5 & -3 \\ 1 & 7 \end{bmatrix} \begin{bmatrix} 2 & -1 \\ 3 & 0 \end{bmatrix} = \begin{bmatrix} 5(2)-3(3) & 5(-1)-3(0) \\ 1(2)+7(3) & 1(-1)+7(0) \end{bmatrix} = \begin{bmatrix} 1 & -5 \\ 23 & -1 \end{bmatrix}$

29. $AB = \begin{bmatrix} 2 & 1 & -1 \\ 4 & 0 & 1 \end{bmatrix} \begin{bmatrix} 2 & -3 \\ 1 & 2 \\ 4 & 8 \end{bmatrix} =$

$$= \begin{bmatrix} 2(2)+1(1)-1(4) & 2(-3)+1(2)-1(8) \\ 4(2)+0(1)+1(4) & 4(-3)+0(2)+1(8) \end{bmatrix} = \begin{bmatrix} 1 & -12 \\ 12 & -4 \end{bmatrix}$$

$BA = \begin{bmatrix} 2 & -3 \\ 1 & 2 \\ 4 & 8 \end{bmatrix} \begin{bmatrix} 2 & 1 & -1 \\ 4 & 0 & 1 \end{bmatrix} =$

$$= \begin{bmatrix} 2(2)-3(4) & 2(1)-3(0) & 2(-1)-3(1) \\ 1(2)+2(4) & 1(1)+2(0) & 1(-1)+2(1) \\ 4(2)+8(4) & 4(1)+8(0) & 4(-1)+8(1) \end{bmatrix} = \begin{bmatrix} -8 & 2 & -5 \\ 10 & 1 & 1 \\ 40 & 4 & 4 \end{bmatrix}$$

31. $\left[\begin{array}{cc|cc} 1 & 2 & 1 & 0 \\ 3 & 2 & 0 & 1 \end{array}\right] \xrightarrow{-3\,①+②} \left[\begin{array}{cc|cc} 1 & 2 & 1 & 0 \\ 0 & -4 & -3 & 1 \end{array}\right] \xrightarrow{-\frac{1}{4}\,③}$

$\left[\begin{array}{cc|cc} 1 & 2 & 1 & 0 \\ 0 & 1 & \frac{3}{4} & -\frac{1}{4} \end{array}\right] \xrightarrow{-2\,②+①} \left[\begin{array}{cc|cc} 1 & 0 & -\frac{1}{2} & \frac{1}{2} \\ 0 & 1 & \frac{3}{4} & -\frac{1}{4} \end{array}\right]$

33. $\left[\begin{array}{cc|cc} -1 & 4 & 1 & 0 \\ 3 & 5 & 0 & 1 \end{array}\right] \xrightarrow{3\,①+②} \left[\begin{array}{cc|cc} -1 & 4 & 1 & 0 \\ 0 & 17 & 3 & 1 \end{array}\right] \begin{array}{c} \xrightarrow{\frac{1}{17}\,②} \\ \xrightarrow{-1\,①} \end{array}$

$\left[\begin{array}{cc|cc} 1 & -4 & -1 & 0 \\ 0 & 1 & \frac{3}{17} & \frac{1}{17} \end{array}\right] \xrightarrow{4\,②+①} \left[\begin{array}{cc|cc} 1 & 0 & -\frac{5}{17} & \frac{4}{17} \\ 0 & 1 & \frac{3}{17} & \frac{1}{17} \end{array}\right]$

35.
$$\left[\begin{array}{ccc|ccc} 1 & 1 & -1 & 1 & 0 & 0 \\ 2 & 0 & 1 & 0 & 1 & 0 \\ 0 & -2 & 3 & 0 & 0 & 1 \end{array}\right] \xrightarrow{\begin{array}{c}-2①+\\②\end{array}} \left[\begin{array}{ccc|ccc} 1 & 1 & -1 & 1 & 0 & 0 \\ 0 & -2 & 3 & -2 & 1 & 0 \\ 0 & -2 & 3 & 0 & 0 & 1 \end{array}\right] \xrightarrow{-②+③}$$

$$\left[\begin{array}{ccc|ccc} 1 & 1 & -1 & 1 & 0 & 0 \\ 0 & -2 & 3 & -2 & 1 & 0 \\ 0 & 0 & 0 & -2 & 1 & 1 \end{array}\right] \quad \text{Inverse does not exist.}$$

37.
$$\left[\begin{array}{cc|cc} 3 & 5 & 1 & 0 \\ -1 & 3 & 0 & 1 \end{array}\right] \xrightarrow{3②+①} \left[\begin{array}{cc|cc} 0 & 14 & 1 & 3 \\ -1 & 3 & 0 & 1 \end{array}\right] \xrightarrow[\frac{1}{14}①]{-②}$$

$$\left[\begin{array}{cc|cc} 0 & 1 & \frac{1}{14} & \frac{3}{14} \\ 1 & -3 & 0 & 1 \end{array}\right] \xrightarrow{3①+②} \left[\begin{array}{cc|cc} 0 & 1 & \frac{1}{14} & \frac{3}{14} \\ 1 & 0 & \frac{3}{14} & -\frac{5}{14} \end{array}\right] \xrightarrow{①\leftarrow②}$$

$$\left[\begin{array}{cc|cc} 1 & 0 & \frac{3}{14} & -\frac{5}{14} \\ 0 & 1 & \frac{1}{14} & \frac{3}{14} \end{array}\right]$$

$$\left[\begin{array}{cc} 3 & 5 \\ -1 & 3 \end{array}\right]\left[\begin{array}{c} x \\ y \end{array}\right] = \left[\begin{array}{c} 13 \\ 5 \end{array}\right], \qquad \left[\begin{array}{cc} \frac{3}{14} & -\frac{5}{14} \\ \frac{1}{14} & \frac{3}{14} \end{array}\right]\left[\begin{array}{c} 13 \\ 5 \end{array}\right] = \left[\begin{array}{c} 1 \\ 2 \end{array}\right]$$

39.
$$\left[\begin{array}{cc|cc} 1 & -2 & 1 & 0 \\ 1 & 2 & 0 & 1 \end{array}\right] \xrightarrow{-①+②} \left[\begin{array}{cc|cc} 1 & -2 & 1 & 0 \\ 0 & 4 & -1 & 1 \end{array}\right] \xrightarrow{\frac{1}{4}②}$$

$$\left[\begin{array}{cc|cc} 1 & -2 & 1 & 0 \\ 0 & 1 & -\frac{1}{4} & \frac{1}{4} \end{array}\right] \xrightarrow{2②+①} \left[\begin{array}{cc|cc} 1 & 0 & \frac{1}{2} & \frac{1}{2} \\ 0 & 1 & -\frac{1}{4} & \frac{1}{4} \end{array}\right]$$

$$\left[\begin{array}{cc} 1 & -2 \\ 1 & 2 \end{array}\right]\left[\begin{array}{c} x \\ y \end{array}\right] = \left[\begin{array}{c} 1 \\ 3 \end{array}\right], \qquad \left[\begin{array}{cc} \frac{1}{2} & \frac{1}{2} \\ -\frac{1}{4} & \frac{1}{4} \end{array}\right]\left[\begin{array}{c} 1 \\ 3 \end{array}\right] = \left[\begin{array}{c} 2 \\ \frac{1}{2} \end{array}\right]$$

41.

$$\left[\begin{array}{rrr|rrr} 1 & -2 & 3 & 1 & 0 & 0 \\ -3 & 4 & -1 & 0 & 1 & 0 \\ 2 & 1 & 4 & 0 & 0 & 1 \end{array}\right] \begin{array}{l} 3①+② \\ \xrightarrow{\hspace{2cm}} \\ -2①+③ \end{array}$$

$$\left[\begin{array}{rrr|rrr} 1 & -2 & 3 & 1 & 0 & 0 \\ 0 & -2 & 8 & 3 & 1 & 0 \\ 0 & 5 & -10 & -2 & 0 & 1 \end{array}\right] \xrightarrow{-\frac{1}{2}②}$$

$$\left[\begin{array}{rrr|rrr} 1 & -2 & 3 & 1 & 0 & 0 \\ 0 & 1 & -4 & -\frac{3}{2} & -\frac{1}{2} & 0 \\ 0 & 5 & -10 & -2 & 0 & 1 \end{array}\right] \begin{array}{l} 2②+① \\ \xrightarrow{\hspace{2cm}} \\ -5②+③ \end{array}$$

$$\left[\begin{array}{rrr|rrr} 1 & 0 & -5 & -2 & -1 & 0 \\ 0 & 1 & -4 & -\frac{3}{2} & -\frac{1}{2} & 0 \\ 0 & 0 & 10 & \frac{11}{2} & \frac{5}{2} & 1 \end{array}\right] \xrightarrow{\frac{1}{10}③}$$

$$\left[\begin{array}{rrr|rrr} 1 & 0 & -5 & -2 & 1 & 0 \\ 0 & 1 & -4 & -\frac{3}{2} & -\frac{1}{2} & 0 \\ 0 & 0 & 1 & \frac{11}{20} & \frac{1}{4} & \frac{1}{10} \end{array}\right] \begin{array}{l} 5③+① \\ \xrightarrow{\hspace{2cm}} \\ 4③+② \end{array}$$

$$\left[\begin{array}{rrr|rrr} 1 & 0 & 0 & \frac{15}{20} & \frac{1}{4} & \frac{1}{2} \\ 0 & 1 & 0 & \frac{7}{20} & \frac{1}{2} & \frac{4}{10} \\ 0 & 0 & 1 & \frac{11}{20} & \frac{1}{4} & \frac{1}{10} \end{array}\right]$$

$$\left[\begin{array}{rrr} 1 & -2 & 3 \\ -3 & 4 & -1 \\ 2 & 1 & 4 \end{array}\right]\left[\begin{array}{c} a \\ b \\ c \end{array}\right] = \left[\begin{array}{r} 4 \\ -2 \\ 3 \end{array}\right], \quad \left[\begin{array}{ccc} \frac{15}{20} & \frac{5}{20} & \frac{10}{20} \\ \frac{14}{20} & \frac{10}{20} & \frac{8}{20} \\ \frac{11}{20} & \frac{5}{20} & \frac{2}{20} \end{array}\right]\left[\begin{array}{r} 4 \\ -2 \\ 3 \end{array}\right] = \left[\begin{array}{c} 4 \\ 3 \\ 2 \end{array}\right]$$

43. $\begin{vmatrix} 2 & -1 \\ 3 & 2 \end{vmatrix} = 4 - (-3) = 7$

45. $\begin{vmatrix} -8 & 6 \\ 5 & -2 \end{vmatrix} = 16 - (30) = -14$

47. $\begin{vmatrix} 7 & -3 & 2 \\ 4 & -5 & 12 \\ -21 & 9 & -6 \end{vmatrix} = 210 + 756 + 72 - (210+756+72) = 0$

49.

$$\begin{vmatrix} 1 & 5 & 4 \\ -2 & 3 & -1 \\ 2 & -1 & 5 \end{vmatrix} = 1 \begin{vmatrix} 3 & -1 \\ -1 & 5 \end{vmatrix} - 5 \begin{vmatrix} -2 & -1 \\ 2 & 5 \end{vmatrix} + 4 \begin{vmatrix} -2 & 3 \\ 2 & -1 \end{vmatrix} =$$

$$= 1(15-1) - 5(-10+2) + 4(2-6) = 38$$

51.

$$\begin{vmatrix} 2 & 2 & -1 & 1 \\ 0 & 1 & 3 & 2 \\ -1 & 0 & 1 & 1 \\ 0 & 2 & -2 & 1 \end{vmatrix} \xrightarrow{2\,③ + ①} \begin{vmatrix} 0 & 2 & 1 & 3 \\ 0 & 1 & 3 & 2 \\ -1 & 0 & 1 & 1 \\ 0 & 2 & -2 & 1 \end{vmatrix} =$$

$$-1 \begin{vmatrix} 2 & 1 & 3 \\ 1 & 3 & 2 \\ 2 & -2 & 1 \end{vmatrix} = -\left[6 + 4 - 6 - (18-8+1) \right] = 7$$

53.

$$\begin{vmatrix} 2 & 5 \\ x & 1 \end{vmatrix} = 0$$

$$2 - 5x = 0$$

$$x = \frac{2}{5}$$

55. $x = \dfrac{\begin{vmatrix} 8 & -2 \\ 19 & -3 \end{vmatrix}}{\begin{vmatrix} 5 & -2 \\ 4 & -3 \end{vmatrix}} = \dfrac{14}{-7} = -2,$ $y = \dfrac{\begin{vmatrix} 5 & 8 \\ 4 & 19 \end{vmatrix}}{-7} = \dfrac{63}{-7} = -9$

57.

$$a = \frac{\begin{vmatrix} 1 & -1 & 3 \\ 3 & 0 & -2 \\ 0 & 0 & 5 \end{vmatrix}}{\begin{vmatrix} 0 & -1 & 3 \\ 1 & 0 & -2 \\ 2 & 0 & 5 \end{vmatrix}} = \frac{1 \begin{vmatrix} 3 & -2 \\ 0 & 5 \end{vmatrix}}{1 \begin{vmatrix} 1 & -2 \\ 2 & 5 \end{vmatrix}} = \frac{15}{9} = \frac{5}{3}$$

$$b = \frac{\begin{vmatrix} 0 & 1 & 3 \\ 1 & 3 & -2 \\ 2 & 0 & 5 \end{vmatrix}}{9} = \frac{-1 \begin{vmatrix} 1 & -2 \\ 2 & 5 \end{vmatrix} + 3 \begin{vmatrix} 1 & 3 \\ 2 & 0 \end{vmatrix}}{9} = \frac{-5 - 22}{2} = -3$$

$$c = \frac{\begin{vmatrix} 0 & -1 & 1 \\ 1 & 0 & 3 \\ 2 & 0 & 0 \end{vmatrix}}{9} = \frac{1 \begin{vmatrix} 1 & 3 \\ 2 & 0 \end{vmatrix}}{9} = \frac{-6}{9} = \frac{-2}{3}$$

59.

$$r = \frac{\begin{vmatrix} 4 & -1 & -1 \\ 2 & 3 & -2 \\ 4 & 6 & -4 \end{vmatrix}}{\begin{vmatrix} 3 & -1 & -1 \\ 1 & 3 & -2 \\ 2 & 6 & -4 \end{vmatrix}} = \frac{48 + 8 - 12 - (-12 - 48 - 8)}{-36 + 4 - 6 - (-6 - 36 + 4)} = \frac{112}{0}, \quad \text{Dependent}$$

61. $x + y = 90$
$\underline{x + 3 = y}$

$x + y = 90$
$\underline{x - y = -3}$

$$2x = 87, \quad x = 43.5$$
$$y = 90 - 43.5 = 46.5$$

63.
$$x + \quad y = \quad 750,000$$
$$.09x + .11y = \quad 80,000$$

$$-.09x - .09y = - \quad 67,000$$
$$.09x + .11y = \quad 80,000$$

$$.02y = \quad 12,500$$
$$y = \quad 625,000 \text{ at } 9\%$$
$$x = \quad 125,000 \text{ at } 11\%$$

$$x \to 10\%, \quad y \to 30\%, \quad z \to 50\%$$

65. $x + y + z = 250$
$$2y = z$$
$$.10x + .30y + .50z = .40(250)$$

$$x + y + 2y = 250$$
$$10x + 30y + 50(2y) = 10,000$$

$$10x + \quad 30y = \quad 2,500$$
$$10x + 130y = 10,000$$

$$100y = \quad 7,500$$
$$y = 75, \quad z = 150$$
$$x = 250 - 75 - 150 = 25$$

71. $x + y = 12$
$$x = 8, \quad y = 4$$
$$p = x + 2y$$

$(0, 12) \quad p = 0 + 2(12) = 24, \text{ maximum}$

$(8, 0) \quad p = 8 + 2(0) = 8$

$(8, 4) \quad p = 8 + 2(4) = 16$

$(0, 0) \quad p = 0$

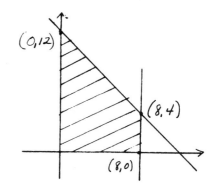

73. $x + \quad y = 4$
$$x + 2y = 6$$
$$-y = -2, \quad y = 2, \quad x = 2$$
$$p = 3x + 5y$$

$(0, 4) \quad p = 3(0) + 5(4) = 20$

$(6, 0) \quad p = 3(6) + 5(0) = 18$

$(2, 2) \quad p = 3(2) + 5(2) = 16, \text{ minimum}$

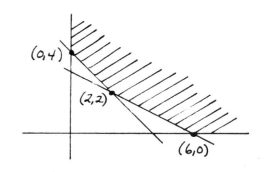

Chapter 7 Vectors and Complex Numbers

1. $|F| = \sqrt{1^2+2^2} = \sqrt{5} = 2.24$
    $\tan^{-1}\theta = 2, \theta = 63.4°$

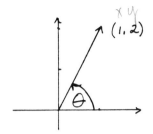

3. $|F| = \sqrt{2+7} = 3$
    $\tan^{-1}\theta = \dfrac{\sqrt{7}}{\sqrt{2}}, \theta = 61.9°$

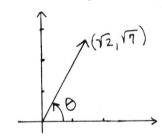

5. $|F| = \sqrt{(-4)^2+(-4)^2} = \sqrt{32} = 5.66$
    $\tan^{-1}\theta = 1, \theta = 135°$

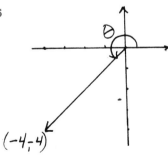

7. $|\mathbf{F}| = \sqrt{4^2+(-3)^2} = \sqrt{25} = 5$

$\tan \theta = \frac{-3}{4}, \theta = -36.9°$

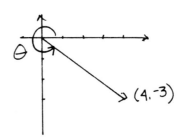

9. $|\mathbf{F}| = \sqrt{3+6^2} = \sqrt{39} = 6.24$

$\tan \theta = \frac{6}{-\sqrt{3}}, \theta = 106.1°$

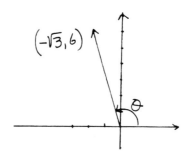

11. $F_x = 10 \cos 50° = 6.43$

$F_y = 10 \sin 50° = 7.66$

13. $F_x = 13.7 \cos 34°10' = 11.34$

$F_y = 13.7 \sin 34°10' = 7.69$

15. $F_x = 158 \cos 125° = -90.63$

$F_y = 158 \sin 125° = 129.43$

17. $F_x = 43.5 \cos 220° = -33.32$

$F_y = 43.5 \sin 220° = -27.96$

19. $F_x = 10.4 \cos 335° = 9.43$

$F_y = 10.4 \sin 335° = -4.40$

21. $|\mathbf{F}| = \sqrt{20^2+15^2} = 25$

$\tan \theta = \frac{15}{20}, \theta = 36.87°$

23. $|\mathbf{F}| = \sqrt{17.5^2+69.3^2} = 71.5$

$\tan \theta = \frac{69.3}{17.5}, \theta = 75.83°$

25. $|\mathbf{F}| = \sqrt{.130^2+.080^2} = .153$

$\tan \theta = \frac{.080}{.130}, \theta = 31.6°$

27. $|\mathbf{F}| = \sqrt{64+49} = 10.6$

$\tan \theta = \frac{-7}{8}, \theta = 318.8°$

29. $|\mathbf{F}| = \sqrt{4+49} = 7.28$

$\tan \theta = \frac{-7}{-2}, \theta = 254.1°$

31. $F_y = 75 \sin 12°10' = 15.8$, vertical

$F_x = 75 \cos 12°10' = 73.3$, horizontal

33. $|\mathbf{V}| = \sqrt{300^2+900^2} = 948.7$

35. $|\mathbf{F}| = \sqrt{3^2 + 5^2} = 5.83$

   $\tan\theta = \frac{5}{3}$, $\theta = 59°04'$ with shore line

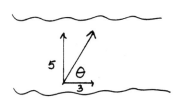

Section 7.2   Operations on Vectors

1. $A+B = (2+3)i + (2-1)j$
      $= 5i+j$

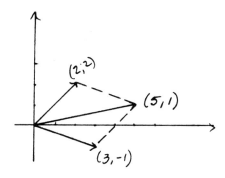

3. $A+B = (-2 - 4)i + (3-5)j = -6i - 2j$

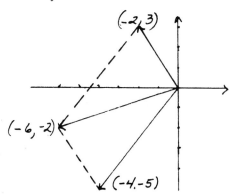

5. $A+B = (-1 + 2)i + (-1 - 2)j = +i - 3j$

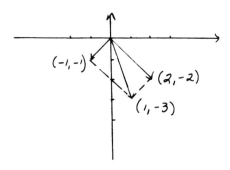

7. $A+B = (-4 + 4)i + (-6 + 2)j = -4j$

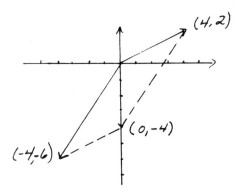

9. $A+B = (5+1)i + (-5 - 2)j = 6i-7j$

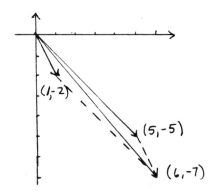

11. $A_x = 16 \cos 25° = 14.50$ $A_y = 16 \sin 25° = 6.76$

    $B_x = 22 \cos 70° = 7.52$ $B_y = 22 \sin 70° = 20.67$

    $R_x = 14.5 + 7.52 = 22.02$ $R_y = 6.76 + 20.67 = 27.43$

    $|R| = \sqrt{22.02^2 + 27.43^2} = 35.18$

    $\tan \theta = \frac{27.43}{22.02}, \theta = 51.2°$

13. $A_x = 9.5 \cos 90° = 0$ $A_y = 9.5 \sin 90° = 9.5$

    $B_x = 5.1 \cos 40° = 3.91$ $B_y = 5.1 \sin 40° = 3.28$

    $R_x = 0 + 3.91 = 3.91$ $R_y = 9.5 + 3.28 = 12.78$

    $|R| = \sqrt{3.91^2 + 12.78^2} = 13.36$

    $\tan \theta = \frac{12.78}{3.91}, \theta = 72.99°$

15. $A_x = 29.2 \cos 15.6° = 28.12$ $A_y = 29.2 \sin 15.6° = 7.85$

    $B_x = 82.6 \cos 150° = -71.53$ $B_y = 82.6 \sin 150° = 41.3$

    $R_x = 28.12 - 71.53 = -43.41$ $R_y = 7.85 + 41.3 = 49.15$

    $|R| = \sqrt{43.41^2 + 49.15^2} = 65.58$

    $\tan \theta = \frac{49.15}{-43.41}, \theta = 131.45°$

17. $A_x = 550 \cos 140° = -421.32$ $A_y = 550 \sin 140° = 353.53$

    $B_x = 925 \cos 310° = 594.58$ $B_y = 925 \sin 310° = -708.59$

    $R_x = -421.32 + 594.58 = 173.26$ $R_y = 353.53 - 708.59 = -355.06$

    $|R| = \sqrt{173.26^2 + 355.06^2} = 395.08$

    $\tan \theta = \frac{-355.06}{173.26}, \theta = -63.99°$

19. $\tan \theta = \frac{300}{50}, \theta = 99.46°$

    $|R| = \sqrt{300^3 + 50^2} = 304.14$

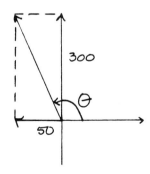

21. $R_x = 200 \cos 35° = 163.83$

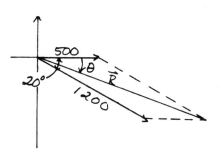

23. $A_x = 60$ $A_y = 0$

     $B_x = 75 \cos 55° = 43.02$ $B_y = 75 \sin 55° = 61.44$

     $R_x = 103.02$ $R_y = 61.44$

     $|\mathbf{R}| = \sqrt{103.02^2 + 61.44^2} = 119.94$

     $\tan \theta = \dfrac{61.44}{103.02}, \ \theta = 30.8°$

25. $|\mathbf{B}| = 1200, \ \theta_B = -20°$

     $|\mathbf{P}| = 500, \ \theta_P = 0°$

     $B_x = 1200 \cos (-20°) = 1127.63$ $B_y = 1200 \sin (-20°) = -410.42$

     $P_x = 500 \cos 0° = 500$ $P_y = 500 \sin 0° = 0$

     $R_x = 1627.63$ $R_y = -410.42$

     $|\mathbf{R}| = \sqrt{1627.63^2 + 410.42^2} = 1679$

     $\tan \theta = \dfrac{-410.42}{1627.63}, \ \theta = -14.2°$

Section 7.3   The Dot Product of Two Vectors

  1. $\mathbf{A} \cdot \mathbf{B} = 1(2) + 1(-1) = 1$ 3. $\mathbf{A} \cdot \mathbf{B} = 6(7) + 1(-1) = 41$

  5. $\mathbf{A} \cdot \mathbf{B} = 1(1) + 0(3) = 1$ 7. $\mathbf{A} \cdot \mathbf{B} = 25(40) \cos (85° - 27°) = 529.9$

9. $\mathbf{A} \cdot \mathbf{B} = 190(75) \cos (205° - 100°) = -3688$

11. $\mathbf{A} \cdot \mathbf{B} = (24.3)(16.2) \cos (276° - 245°) = 337.4$

13. $\mathbf{A} \cdot \mathbf{B} = 10 - 1 = 9, |\mathbf{A}| = \sqrt{5^2 + 1} = \sqrt{26}, |\mathbf{B}| = \sqrt{2^2 + 1} = \sqrt{5}$

   $\cos \theta = \dfrac{\mathbf{A} \cdot \mathbf{B}}{|\mathbf{A}||\mathbf{B}|} = \dfrac{9}{\sqrt{5}\sqrt{26}}, \theta = 37.87°$

15. $\mathbf{A} \cdot \mathbf{B} = -6 - 12 = -18, |\mathbf{A}| = \sqrt{9 + 16} = 5, |\mathbf{B}| = \sqrt{4 + 9} = \sqrt{13}$

   $\cos \theta = \dfrac{-18}{5\sqrt{13}}, \theta = 176.82°$

17. $W = |20||30| \cos 28° = 529.8$ 19. $W = |5||100| \cos 30° = 433$

Section 7.4   Complex Numbers

7. $i^{101} = i^{4(25)+1} = (i^4)^{25} i = i$

9. $i^{402} = i^{4(100)+2} = (i^4)^{100} i^2 = i^2 = -i$

17. $x^2 + iy = 9 - 3i$ 19. $3x - y + 7 + i(x + y - 1) = 0$

   $(x^2 - 9) + (y + 3)i = 0$ $3x - y + 7 = 0$

   $x^2 - 9 = 0 \qquad y = -3$ $\underline{x + y - 1 = 0}$

   $x = \pm 3$ $4x \quad + 6 = 0$

                              $x = -\dfrac{3}{2}$

                              $y = 1 - x = \dfrac{5}{2}$

21. $(3 + 2i) + (4 + 3i) = (3 + 4) + (2 + 3)i = 7 + 5i$

23. $(5 - 2i) + (-7 + 5i) = (5 - 7) + (-2 + 5i) = -2 + 3i$

25. $(1 + i) + (3 - i) = (1 + 3) + (1 - 1)i = 4$

27. $(3 + 5i) - 4i = (3 + 0) + (5 - 4)i = 3 + i$

29. $(2+3i)(4+5i) = 8 + 10i + 12i + 15i^2 = -7 + 22i$

31. $(5-i)(5+i) = 25-i^2 = 26$

33. $(4+\sqrt{3}i)^2 = 16 + 8\sqrt{3}i + 3i^2 = 13 + 8\sqrt{3}i$

35. $6i(4-3i) = 24i - 18i^2 = 18+24i$

37. $\dfrac{3+2i}{1+i} \cdot \dfrac{1-i}{1-i} = \dfrac{3-3i+2i-2i^2}{1-i^2} = \dfrac{5-i}{2}$

39. $\dfrac{3}{2-3i} \cdot \dfrac{2+3i}{2+3i} = \dfrac{6+9i}{4-9i^2} = \dfrac{6+9i}{13}$

41. $\dfrac{1}{5i} \cdot \dfrac{-5i}{-5i} = \dfrac{-5i}{-25i^2} = \dfrac{-5i}{25} = \dfrac{-i}{5}$

43. $\dfrac{-1-3i}{4-\sqrt{2}i} \cdot \dfrac{4+\sqrt{2}i}{4+\sqrt{2}\,i} = \dfrac{-4-\sqrt{2}i-12i-3\sqrt{2}i^2}{16-2i^2} = \dfrac{(-4+3\sqrt{2})-(12+\sqrt{2}i)}{18}$

45. $(a+bi)(a-bi) = a^2 - bi^2 = a^2 + b^2$

47. $a+bi = \overline{a+bi}$
$a+bi = a-bi$
$0+2bi = 0$
$\quad 2b = 0,\ b = 0$

49. $\dfrac{z+\bar{z}}{2} = \dfrac{(x+iy)+(x-iy)}{2} = \dfrac{2x}{2} = x$

$\dfrac{z-\bar{z}}{2i} = \dfrac{(x+iy)-(x-iy)}{2i} = \dfrac{2iy}{2i} = y$

1. $(4+i) + (3+5i) = 7+6i$

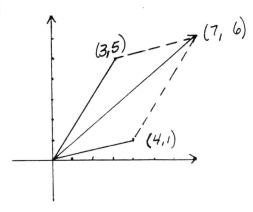

3. $(4+3i) + (-2 + i) = 2+4i$

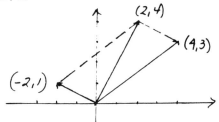

5. $(2-4i) + (-3 + i) = -1 - 3i$

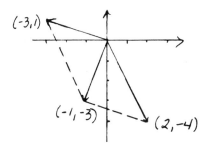

7. $(5+3i) - 6 = -1 + 3i$

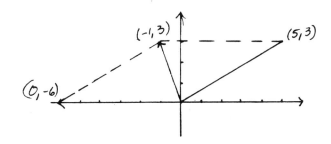

9. $(5-3i) + (5+3i) = 10$

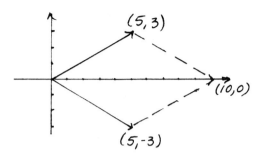

11. $(1+3i) - (2-5i) = -1 + 8i$

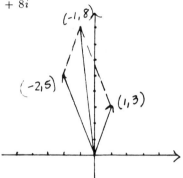

13. $(2+\sqrt{3}i) - (-1-i) = 3 + (\sqrt{3}+i)i$

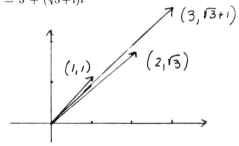

21. $r = \sqrt{1+3} = 2$, $\tan\theta = -\sqrt{3}$, $\theta = -60°$

$z = 2 \operatorname{cis}(-60°)$

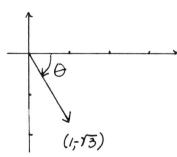

23. $r = \sqrt{5+4} = 3$, $\tan \theta = \frac{2}{\sqrt{5}}$, $\theta = 41.8°$

$z = 3 \text{ cis } 41.8°$

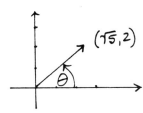

27. $r = \sqrt{9+16} = 5$; $\tan\theta = \frac{-4}{3}$, $\theta = -53.1°$

$z = \text{cis}\,(-53.1°)$

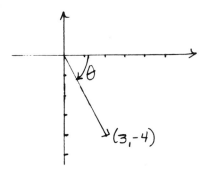

29. $r = \sqrt{25+36} = \sqrt{61}$, $\tan \theta = \frac{-6}{5}$, $\theta = -50.2°$

$z = \sqrt{61} \text{ cis } (-50.2°)$

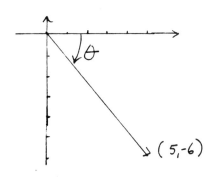

31. $2 \text{ cis } 30° = 2 \cos 30° + 2i \sin 30°$

$= 2\frac{\sqrt{3}}{2} + i2 \cdot \frac{1}{2} = \sqrt{3} + i$

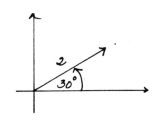

33. $5 \text{ cis } 135° = 5 \cos 135° + 5i \sin 135°$

$$= 5\left(-\frac{\sqrt{2}}{2}\right) + i5\left(\frac{\sqrt{2}}{2}\right) = \frac{5}{2}\sqrt{2}(-1 + i)$$

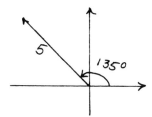

35. $\sqrt{3} \text{ cis } 210° = \sqrt{3} \cos 210° + \sqrt{3}i \sin 210°$

$$= \sqrt{3}\left(-\frac{\sqrt{3}}{2}\right) + \sqrt{3}\left(-\frac{1}{2}\right)i = \frac{-3}{2} - \frac{\sqrt{3}}{2}i$$

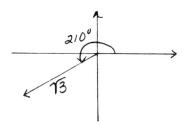

37. $3 \text{ cis } 300° = 3 \cos 300° + 3i \sin 300°$

$$= 3\left(\frac{1}{2}\right) + i(3)\left(\frac{-\sqrt{3}}{2}\right) = \frac{3}{2} - \frac{3\sqrt{3}}{2}i$$

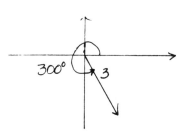

39. $10 \text{ cis } 20° = 10 \cos 20° + 10i \sin 20°$

$$= 9.397 + 3.420i$$

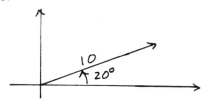

41. $(4 \text{ cis } 30°)(3 \text{ cis } 60°) = 3 \cdot 4 \text{ cis } (30°+60°) = 12 \text{ cis } 90°$

43. $(\sqrt{2} \text{ cis } 90°)(\sqrt{2} \text{ cis } 240°) = (\sqrt{2})^2 \text{ cis } (90° + 240°) = 2 \text{ cis } 330°$

45. $(10 \text{ cis } 35°)(2 \text{ cis } 100°) = 10 \cdot 2 \text{ cis } (35° + 100°) = 20 \text{ cis } 135°$

47. $3 + 4i$, $r = \sqrt{9 + 16} = 5$, $\tan \theta = \frac{4}{3}$, $\theta = 53.1°$

$$3 + 4i = 5 \text{ cis } 53.1°$$

$\sqrt{3} - i$, $r = \sqrt{3 + 1} = 2$, $\tan \theta = \frac{-1}{\sqrt{3}}$, $\theta = -30°$

$$\sqrt{3} - i = 2 \text{ cis } (-30°)$$

$$(3 + 4i)(\sqrt{3} - i) = (5 \text{ cis } 53.1°)\left(2 \text{ cis } (-30°)\right) = 10 \text{ cis } 23.1°$$

49. $\dfrac{10 \text{ cis } 30°}{2 \text{ cis } 90°} = \dfrac{10}{2} \text{ cis } (30° - 90°) = 5 \text{ cis } (-60°)$

51. $\dfrac{4 \text{ cis } 26°40'}{2 \text{ cis } 19°10'} = \dfrac{4}{2} \text{ cis } (26°40' - 19°10') = 2 \text{ cis } 7°30'$

53. $1 - i$, $r = \sqrt{1 + 1} = \sqrt{2}$, $\tan \theta = -1$, $\theta = -45°$

$$1 - i = \sqrt{2} \text{ cis } (-45°)$$

$\sqrt{3} + i$, $r = \sqrt{3 + 1} = 2$, $\tan \theta = \frac{-1}{\sqrt{3}}$, $\theta = 30°$

$$\sqrt{3} + i = 2 \text{ cis } 30°$$

$$\frac{1 - i}{\sqrt{3} + i} = \frac{\sqrt{2} \text{ cis } (-45°)}{2 \text{ cis } 30°} = \frac{\sqrt{2}}{2} \text{ cis } (-45° - 30°) = \frac{\sqrt{2}}{2} \text{ cis } (-75°)$$

55. $4i$, $r = 4$, $\theta = 90°$, $4i = 4 \text{ cis } 90°$

$-1 + i$, $r = \sqrt{1 + 1} = \sqrt{2}$, $\tan \theta = -1$, $\theta = 135°$

$$-1 + i = \sqrt{2} \text{ cis } 135°$$

$$\frac{4i}{-1 + i} = \frac{4 \text{ cis } 90°}{\sqrt{2} \text{ cis } 135°} = \frac{4}{\sqrt{2}} \text{ cis } (90° - 135°) = \frac{4}{\sqrt{2}} \text{ cis } (-45°)$$

57. $\cos \theta$

$$\frac{1}{2}\left(\cos \theta + \text{cis } (-\theta)\right) = \frac{1}{2}\left(\cos \theta + i \sin \theta + \cos (-\theta) + i \sin (-\theta)\right)$$

$$= \frac{1}{2}(\cos \theta + i \sin \theta + \cos \theta - i \sin \theta)$$

$$= \frac{1}{2}(2 \cos \theta) = \cos \theta$$

$$\frac{1}{2i}\left(\text{cis } \theta - \text{cis } (-\theta)\right) = \frac{1}{2i}\left(\cos \theta + i \sin \theta - \cos (-\theta) - i \sin (-\theta)\right)$$

$$= \frac{1}{2i}(\cos \theta + i \sin \theta - \cos \theta + i \sin \theta)$$

$$= \frac{1}{2i}(2i \sin \theta) = \sin \theta$$

1. $(-1 + \sqrt{3}i)^3$, $r = \sqrt{1+3} = 2$, $\tan \theta = -\sqrt{3}$, $\theta = 120°$

   $(-1 + \sqrt{3}i)^3 = \left(2 \text{ cis } (120°)\right)^3 = 2^3 \text{ cis } (360°) = 8 \text{ cis } 0°$

3. $(\sqrt{3} \text{ cis } 60°)^4 = (\sqrt{3})^4 \text{ cis } 4(60°) = 9 \text{ cis } 240°$

5. $(-2 + 2i)^5$, $r = \sqrt{4+4} = \sqrt{8}$, $\tan \theta = \dfrac{2}{-2} = -1$, $\theta = 135°$

   $(-2 + 2i)^5 = (\sqrt{8} \text{ cis } 135°)^5 = (\sqrt{8})^5 \text{ cis } 5(135°) = 128\sqrt{2} \text{ cis } 675°$
   $$= 128\sqrt{2} \text{ cis } 315°$$

7. $(-\sqrt{3} + i)^7$, $r = \sqrt{3+1} = 2$, $\tan \theta = \dfrac{-1}{\sqrt{3}}$, $\theta = 150°$

   $(-\sqrt{3} + i)^7 = \left(2 \text{ cis } (150°)\right)^7 = 2^7 \text{ cis } 7 \,(150°) = 128 \text{ cis } 1050°$
   $$= 128 \text{ cis } 330°$$

9. $(2+5i)^4$, $r = \sqrt{4+25} = \sqrt{29}$, $\tan \theta = \dfrac{5}{2}$, $\theta = 68.2°$

   $(2+5i)^4 = (\sqrt{29} \text{ cis } 68.2°)^4 = 29^2 \text{ cis } 4(68.2°) = 841 \text{ cis } 272.8°$

11. $(1 \text{ cis } 0°)^{1/5} = 1^{1/5} \text{ cis } \left(\dfrac{0 + k360°}{5}\right)$
   $$= \text{cis } 0°, \text{ cis } 72°, \text{ cis } 144°, \text{ cis } 216°, \text{ cis } 288°$$

13. $i^{1/4} = (1 \text{ cis } 90°)^{1/4} = \text{cis } \dfrac{(90° + k360°)}{4}$
   $$= \text{cis } 22.5°, \text{ cis } 112.5°, \text{ cis } 202.5°, \text{ cis } 292.5°$$

15. $(1+i)^{1/2} = (\sqrt{2} \text{ cis } 45°)^{1/2} = 2^{1/4} \text{ cis } \dfrac{(45° + k360°)}{2}$
   $$= \sqrt[4]{2} \text{ cis } 22.5°, \ \sqrt[4]{2} \text{ cis } 202.5°$$

17. $(-\sqrt{3} + i)^{1/6}$, $r = \sqrt{3+1} = 2$, $\tan \theta = \frac{-1}{\sqrt{3}}$, $\theta = 150°$

$$(-\sqrt{3} + i)^{1/6} = (2 \text{ cis } 150°)^{1/6} = 2^{1/6} \text{ cis } \frac{(150 + k360°)}{6}$$

$$= \sqrt[6]{2} \text{ cis } (250° + M60°), \; M = 0, 1, 2, 3, 4, 5, 6$$

19. $(-1 + 3i)^{1/4}$, $r = \sqrt{1+3} = 2$, $\tan \theta = -\sqrt{3}$, $\theta = 120°$

$$(-1 + 3i)^{1/4} = (2 \text{ cis } 120°)^{1/4} = 2^{1/4} \text{ cis } \frac{(120° + k360°)}{4}$$

$$= 2^{1/4} \text{ cis } 30°, \; 2^{1/4} \text{ cis } 120°, \; 2^{1/4} \text{ cis } 210° \; 2^{1/4} \text{ cis } 300°$$

21. $(\text{cis } \theta)^2 = 1 \cos 2\theta$

$$\sin 2\theta = \sqrt{1 - \cos^2 2\theta} = \sqrt{1 - \left((\text{cis } \theta)^2\right)^2} = \sqrt{1 - (\text{cis } \theta)^4}$$

23. $x = (-64)^{1/3}$ $\quad r = 64$, $\theta = 180°$, $-64 = 64 \text{ cis } 180°$

$$(-64)^{1/3} = (64 \text{ cis } 180°)^{1/3} = 64^{1/3} \text{ cis } \frac{(180° + k360°)}{3}$$

$$= 4 \text{ cis } 60°, \; 4 \text{ cis } 180°, \; 4 \text{ cis } 300°$$

Chapter 7   Review

1. $|\text{F}| = \sqrt{9+25} = \sqrt{34}$, $\tan \theta = \frac{5}{-3}$, $\theta = 120.96°$

3. $|\text{F}| = \sqrt{\pi^2 + (5.4)^2} = 6.25$, $\tan \theta = \frac{-5.4}{\pi}$, $\theta = -59.81°$

5. $|\text{F}| = \sqrt{\pi + 2} = 2.268$, $\tan \theta = \frac{-\sqrt{\pi}}{\sqrt{2}}$, $\theta = -51.4°$

7. $x = 15.6 \cos (-134°) = -10.84$
   $y = 15.6 \sin (-134°) = -11.22$

9. $x = 1000 \cos 2.65 = -881.58$

     $y = 1000 \sin 2.65 = 472.03$

11. $A+B = (8i + 4j) + (-2i + j) = (8-2)i + (4+1)j = 6i + 5j$

13. $(i - 3j) + (2i + 3j) = (1 + 2)i + (-3 + 3)j = 3i$

15. $(-i - 2j) + (i - 2j) = (-1 + 1)i + (-2 - 2)j = -4j$

17. $A_x = 100 \cos 30° = 86.6$ $B_x = 83.2 \cos 41.5° = 62.3$

     $A_y = 100 \sin 30° = 50$ $B_y = 83.2 \sin 41.5° = 55.1$

     $A_x + B_x = 86.6 + 62.3 = 148.9$

     $A_y + B_y = 50 + 55.1 = 105.1$

     $A+B = 148.9i + 105.1j$

17. $A_x = .5 \cos (-.34) = .471$ $B_x = 1.3 \cos 1.6 = -.038$

     $A_y = .5 \sin (-.34) = -.167$ $B_y = 1.3 \sin (1.6) = 1.299$

     $A_x + B_x = .471 - .038 = .433$

     $A_y + B_y = -.167 + 1.299 = 1.133$

     $A+B = .471i + 1.133j$

21. $A \cdot B = 8(-2) + 4(1) = -12$

     $|A| = \sqrt{8^2 + 4^2} = \sqrt{80}, |B| = \sqrt{2^2 + 1^2} = \sqrt{5}$

     $\cos \theta = \dfrac{-12}{\sqrt{5}\sqrt{80}}, \theta = 126.9$

23. $A \cdot B = 1(2) - 3(3) = -7$

     $|A| = \sqrt{1^2 + 3^2} = \sqrt{10}, |B| = \sqrt{2^2 + 3^2} = \sqrt{13}$

     $\cos \theta = \dfrac{-7}{\sqrt{10}\sqrt{13}}, \theta = 127.9°$

25. $\mathbf{A} \cdot \mathbf{B} = 1(-1) + (-2)(-2) = 3$

$|\mathbf{A}| = \sqrt{1^2 + 2^2} = \sqrt{5}, |\mathbf{B}| = \sqrt{1^2 + 2^2} = \sqrt{5}$

$\cos \theta = \dfrac{3}{\sqrt{5}\sqrt{5}} = \dfrac{3}{5}, \theta = 53.1°$

27. $\mathbf{A} \cdot \mathbf{B} = (100)(83.2) \cos(41.5° - 30°) = 8320 \cos(11.5°) = 8152.9$

29. $\mathbf{A} \cdot \mathbf{B} = (.5)(1.3) \cos(1.6 + 3.4) = .065 \cos 1.94 = -.235$

$1.94 = 111.2°$

31. $(3+4i) - (5-2i) = (3-5) + (4+2)i = -2 + 6i$

$r = \sqrt{2^2 + 6^2} = \sqrt{40}, \tan \theta = \dfrac{6}{-2}, \theta = 108.4°$

$-2 + 6i = \sqrt{40} \text{ cis } 108.4°$

33. $(6+3i) - (\overline{6+3i}) = (6+3i) - (6-3i) = 6i$

$r = 6, \theta = 90°$

$6i = 6 \text{ cis } 90°$

35. $(2+i)(i-1) = 2i - 2 + i^2 - i = -3 + i$

$r = \sqrt{3^2 + 1^2} = \sqrt{10}, \tan \theta = \dfrac{-1}{3}, \theta = 161.6°$

$-3 + i = \sqrt{10} \text{ cis } 161.6°$

37. $(2+3i)(4+i) = 8 + 2i + 12i + 3i^2 = 5 + 14i$

$r = \sqrt{5^2 + 14^2} = \sqrt{221}, \tan \theta = \dfrac{14}{5}, \theta = 70.3°$

$5 + 14i = \sqrt{221} \text{ cis } 70.3°$

39.

$\text{cis } .45 = \cos .45 + i \sin .45 = .8984 + .4350i$

$2 \text{ cis }(-.95) = 2 \cos(-.95) + 2i \sin(-.95) = 1.163 - 1.627i$

$\text{cis } .45 + 2 \text{ cis }(-.95) = (.8984 + 1.163) + (.4350 - 1.627)i$

$= 2.061 - 1.192i$

$r = \sqrt{2.061^2 + 1.192^2} = 2.38$

$\tan \theta = \dfrac{-1.192}{2.061}, \theta = -30.1°$

$2.061 - 1.192i = 2.38 \text{ cis }(-30.1°)$

41. $\dfrac{4 \cos 4.6}{\mathrm{cis}\,(-2.9)} = 4\,\mathrm{cis}\,(4.6 + 2.9) = 4\,\mathrm{cis}\,7.5 = 4\,\mathrm{cis}\,69.72°$

$$= 4(\cos 69.72° + i \sin 69.72°) = 1.39 + 3.75i$$

43. $\dfrac{2+3i}{4+i} \cdot \dfrac{4-i}{4-i} = \dfrac{8-2i+12i-3i^2}{16-i^2} = \dfrac{11}{17} + \dfrac{10i}{17}$

$r = \sqrt{\left(\dfrac{11}{17}\right)^2 + \left(\dfrac{10}{17}\right)^2} = \dfrac{\sqrt{221}}{17},\ \tan\theta = \dfrac{\frac{10}{17}}{\frac{11}{17}} = \dfrac{10}{11},\ \theta = 42.3°$

$\dfrac{11}{17} + \dfrac{10i}{17} = \dfrac{\sqrt{221}}{17}\,\mathrm{cis}\,42.3°$

45. $(\sqrt{5}\,\mathrm{cis}\,120°)(7\,\mathrm{cis}\,80°) = 7\sqrt{5}\,\mathrm{cis}\,(120° + 80°) = 7\sqrt{5}\,\mathrm{cis}\,200°$

$7\sqrt{5}\,(\cos 200° + i \sin 200°) = -14.7 - 5.35i$

47. $100\left[\mathrm{cis}\,(-120\pi) + \mathrm{cis}\,(120\pi)\right]$

$= 100\left[\cos(-120\pi) + i \sin(-120\pi) + \cos(120\pi) + i \sin(120\pi)\right]$

$= 100\left[\cos(-120\pi) - i \sin(120\pi) + \cos(120\pi) + i \sin(120\pi)\right]$

$= 200 \cos 120\pi = 200 \cos 0° = 200$

49. $3.4 + 4i,\ r = \sqrt{3.4^2 + 4^2} = 27.56,\ \tan\theta = \dfrac{4}{3.4},\ \theta = 49.6°$

$3.4 + 4i,\ r = 27.56,\ \tan\theta = \dfrac{-4}{3.4},\ \theta = -49.6$

$\dfrac{3.4 + 4i}{3.4 - 4i} = \dfrac{27.56\,\mathrm{cis}\,49.6°}{27.56\,\mathrm{cis}\,(-49.6°)} = \mathrm{cis}\,99.2°$

$= \cos 99.2° + i \sin 99.2°$

$= -.16 + .99i$

51. $-1 = 1\,\mathrm{cis}\,180°$

$(-1)^{1/3} = (\mathrm{cis}\,180°)^{1/3} = \mathrm{cis}\,\dfrac{(180° + k360°)}{3} = \mathrm{cis}\,60°,\ \mathrm{cis}\,180°,\ \mathrm{cis}\,300°$

53. $i = 1\,\mathrm{cis}\,90°$

$i^{1/3} = (\mathrm{cis}\,90°)^{1/3} = \mathrm{cis}\left(\dfrac{90° + k360°}{3}\right) = \mathrm{cis}\,30°,\ \mathrm{cis}\,150°,\ \mathrm{cis}\,270°$

55. $i-1$, $r = \sqrt{1+1} = \sqrt{2}$, $\tan \theta = -1$, $\theta = 135°$

$$(i-1)^{1/3} = (\sqrt{2} \text{ cis } 135°)^{1/3} = 2^{1/6} \text{ cis } \frac{(135° + k360°)}{3}$$

$$= 2^{1/6} \text{ cis } 45°, \ 2^{1/6} \text{ cis } 165°, \ 2^{1/6} \text{ cis } 285°$$

57. $1 + 3i$, $r = \sqrt{1+3^2} = \sqrt{10}$, $\tan \theta = \frac{3}{1}$, $\theta = 71.6°$

$$(1 + 3i)^5 = (\sqrt{10} \text{ cis } 71.6°)^5 = 10^{5/2} \text{ cis } 5(71.6)° = 10^{5/2} \text{ cis } 357.8°$$

59. $|F| = \sqrt{3.4^2 + 2.5^2} = 4.22$

$\tan \theta = \frac{-2.5}{3.4}$, $\theta = -36.3°$

61. $|A| = \sqrt{5^2+3^2} = \sqrt{34}$, $|B| = \sqrt{1^2+3^2} = \sqrt{10}$

$A+B = -5 - 9 = -14$

$\cos \theta = \frac{-14}{\sqrt{34}\sqrt{10}}$, $\theta = 139.4°$

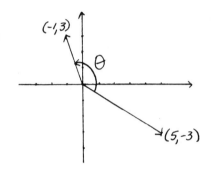

63a. $t^2 + t - 6 = 0$

$(t+3)(t-2) = 0$, $t = 2, -3$

63b. $t^2 - 3t + 2 = 0$

$(t-2)(t-1) = 0$, $t = 1, 2$

63c. $t = 2$

65. $5-4i$, $r = \sqrt{25+16} = \sqrt{41}$, $\tan \theta = \frac{-4}{5}$, $\theta = -38.7°$

$3-i$, $r = \sqrt{9+1} = \sqrt{10}$, $\tan \theta = \frac{-1}{3}$, $\theta = -18.4°$

$$\frac{5-4i}{3-i} = \frac{\sqrt{4} \text{ cis } (-38.7°)}{\sqrt{10} \text{ cis } (-18.4°)} = 2.02 \text{ cis } (-38.7° + 18.4°) = 2.02 \text{ cis } (-20.3°)$$

$$= 2.02 \left(\cos (-20.3°) + i \sin (-20.3°) \right)$$

$$= 1.9 - .7i$$

Chapter 8   The Algebra of Polynomials

<u>Sections 8.1 and 8.2</u>   Polynomials

<u>The Remainder and Factor Theorems</u>

1.
$$
\begin{array}{r}
x\ \ -\ 5 \\
x-1\overline{\smash{)}\ x^2\ -\ 6x\ +\ 2} \\
\underline{x^2\ -\ x} \\
-\ 5x\ +\ 2 \\
\underline{-\ 5x\ +\ 5} \\
-\ 3
\end{array}
$$

$P(1) = 1 - 6 + 2 = -3$

3.
$$
\begin{array}{r}
x^2\ +\ x\ \ -\ 4 \\
x+1\overline{\smash{)}\ x^3\ +\ 2x^2\ -\ 3x\ +\ 4} \\
\underline{x^3\ +\ x^2} \\
x^2\ -\ 3x \\
\underline{x^2\ +\ x} \\
-\ 4x\ +\ 4 \\
\underline{-\ 4x\ -\ 4} \\
8
\end{array}
$$

$P(-1) = (-1)^3 + 2(-1)^2 - 3(-1) + 4 = 8$

5.
$$
\begin{array}{r}
2x^2\ -\ 8x\ \ +\ 17 \\
x+2\overline{\smash{)}\ 2x^3\ -\ 4x^2\ +\ \ \ x\ -\ \ 1} \\
\underline{2x^3\ +\ 4x^2} \\
-\ 8x^2\ +\ \ \ x \\
\underline{-\ 8x^2\ -\ 16x} \\
17x\ -\ \ 1 \\
\underline{17x\ +\ 34} \\
-\ 35
\end{array}
$$

$P(-2) = 2(-2)^3 - 4(-2)^2 + (-2) - 1 = -35$

7.

$$
\begin{array}{r}
3x^3 \;-\; 6x^2 \;+\; 8x \;-\; 16 \\
x + 1 \overline{)\; 3x^4 \;-\; 3x^3 \;+\; 2x^2 \;-\; 8x \;+\; 1} \\
\underline{3x^4 \;+\; 3x^3} \\
-6x^3 \;+\; 2x^2 \\
\underline{-6x^3 \;-\; 6x^2} \\
8x^2 \;-\; 8x \\
\underline{8x^2 \;+\; 8x} \\
-16x \;+\; 1 \\
\underline{-16x \;-\; 16} \\
17
\end{array}
$$

$P(-1) = 3(-1)^3 - 6(-1)^2 + 8(-1) - 16 = 17$

$-3 - 6 - 8 - 16 = 17$

9.

$$
\begin{array}{r}
x^3 \;-\; 2x^2 \;+\; 2x \;-\; 1 \\
x + 2 \overline{)\; x^4 \qquad\quad -\; 2x^2 \;+\; 3x \;-\; 2} \\
\underline{x^4 \;+\; 2x^3} \\
-2x^3 \;-\; 2x^2 \\
\underline{-2x^3 \;-\; 4x^2} \\
2x^2 \;+\; 3x \\
\underline{2x^2 \;+\; 4x} \\
-\; x \;-\; 2 \\
\underline{-\; x \;-\; 2} \\
0
\end{array}
$$

$P(-2) = (-2)^4 - 2(-2)^2 + 3(-2) - 2 = 0$

11.

$$
\begin{array}{r}
2x^2 \;-\; 3x \;+\; 2 \\
x - 3 \overline{)\; 2x^3 \;-\; 9x^2 \;+\; 11x \;-\; 6} \\
\underline{2x^3 \;-\; 6x^2} \\
-3x^2 \;+\; 11x \\
\underline{-3x^2 \;+\; 9x} \\
2x \;-\; 6 \\
\underline{2x \;-\; 6} \\
0
\end{array}
$$

$P(3) = 2(3)^3 - 9(3)^2 + 11(3) - 6 = 0$

13.
$$
\begin{array}{r}
-x^2 + 4x - 2 \\
x+1{\overline{\smash{\big)}\,-x^3 + 3x^2 + 2x - 7}} \\
\underline{-x^3 - x^2} \\
4x^2 + 2x \\
\underline{4x^2 + 4x} \\
-2x - 7 \\
\underline{-2x - 2} \\
-5
\end{array}
$$

$$P(-1) = -(-1)^3 + 3(-1)^2 + 2(-1) - 7 = -5$$

15.
$$
\begin{array}{r}
x^2 + 7x + 10 \\
3x-2{\overline{\smash{\big)}\,3x^3 + 19x^2 + 16x - 20}} \\
\underline{3x^3 - 2x^2} \\
21x^2 + 16x \\
\underline{21x^2 - 14x} \\
30x - 20 \\
\underline{30x - 20} \\
\bullet
\end{array}
$$

$$P\left(\frac{2}{3}\right) = 3\left(\frac{2}{3}\right)^3 + 19\left(\frac{2}{3}\right)^2 + 16\left(\frac{2}{3}\right) - 20 = 0$$

17.
$$
\begin{array}{r}
3x^2 - 10 \\
x^2+1{\overline{\smash{\big)}\,3x^4 - 7x^2 + 1}} \\
\underline{3x^4 + 3x^2} \\
-10x^2 + 1 \\
\underline{-10x^2 - 10} \\
11
\end{array}
$$

19.
$$
\begin{array}{r}
2x - 1 \\
x^2-3x+3{\overline{\smash{\big)}\,2x^3 - 7x^2 + 9x - 3}} \\
\underline{2x^3 - 6x^2 + 6x} \\
-x^2 + 3x - 3 \\
\underline{-x^2 + 3x - 3}
\end{array}
$$

21. $P(4) = 4^2 - 5(4) + 4 = 0,$ Yes

23. $P(-1) = (-1)^3 - 1 = -2,$ No

25. $P(2) = 4(2)^3 + (2)^2 - 16(2) - 4 = 0,$ Yes

27. $P(2) = 3(2)^5 + 7(2) - 8 = 102,$ No

29. $P(4) = 4^4 + 2(4)^3 - 15(4)^2 - 32(4) - 16 = 0,$ Yes

31. $P\left(\frac{1}{3}\right) = 3\left(\frac{1}{3}\right)^3 + 2\left(\frac{1}{3}\right)^2 - 4\left(\frac{1}{3}\right) + 1 = 0,$ Yes

33. $P(-2i) = (-2i)^4 - 16 = 0,$ Yes

35. $P(2i) = (2i)^3 - 2(2i)^2 + 4(2i) - 8 = 0,$ Yes

37. $a(x-1)(x+1)(x-2) = P(x)$

$\quad a(-1)(1)(-2) = -2 = P(0)$

$\quad\quad\quad a = -1$

$\quad -1(x-1)(x+1)(x-2) = P(x)$

39. $a(x-1)^2(x-2)^2 = P(x)$

$\quad P(0) = a(1)(4) = 3$

$\quad\quad\quad a = \frac{3}{4}$

$\quad \frac{3}{4}(x-1)^2(x-2)^2 = P(x)$

41. $a(x-2+i)(x-2+i)(x-\sqrt{2}) = P(x)$

$\quad P(0) = a(-2+i)(-2-i)(-\sqrt{2}) = 2$

$\quad\quad -5\sqrt{2}a = 2, \quad a = -\frac{\sqrt{2}}{5}$

$\quad P(x) = -\frac{\sqrt{2}}{5}(x^2+5)(x-\sqrt{2})$

43. $\begin{vmatrix} -\lambda & -\frac{3}{5} & 0 \\ \frac{5}{3} & -\lambda & -\frac{5}{3} \\ 0 & 6 & -(\lambda+6) \end{vmatrix} = -\lambda \begin{vmatrix} -\lambda & -\frac{5}{3} \\ 6 & -(\lambda+6) \end{vmatrix} - \frac{5}{3} \begin{vmatrix} -\frac{3}{5} & 0 \\ 6 & -(\lambda+6) \end{vmatrix}$

$\quad\quad = -\lambda\left(\lambda(\lambda+6)+10\right) - \frac{5}{3}\left(\frac{3}{5}(\lambda+6)\right)$

$\quad\quad = -\lambda^3 - 6\lambda^2 - 10\lambda - \lambda - 6 = -\lambda^3 - 6\lambda^2 - 11\lambda - 6$

$\quad P(-1) = -(-1)^3 - 6(-1)^2 - 11(-1) - 6 = 0$

1.

$$
4 \,) \, \begin{array}{rrr} 1 & -5 & 4 \\ & 4 & -4 \\ \hline 1 & -1 \end{array}
$$

$$x - 1$$

3.

$$
2 \,) \, \begin{array}{rrrr} 1 & -2 & 3 & -1 \\ & 2 & 0 & 6 \\ \hline 1 & 0 & 3 & 5 \end{array}
$$

$$x^2 + 3 + \frac{5}{x - 2}$$

5.

$$
-\tfrac{1}{2} \,) \, \begin{array}{rrrr} 1 & 0 & -3 & 9 \\ & -\tfrac{1}{2} & \tfrac{1}{4} & \tfrac{11}{8} \\ \hline 1 & -\tfrac{1}{2} & -\tfrac{11}{4} & \tfrac{83}{8} \end{array}
$$

$$x^2 - \frac{x}{2} - \frac{11}{4} + \frac{\frac{83}{8}}{x - \frac{1}{2}}$$

7.

$$
3 \,) \, \begin{array}{rrrrr} 1 & -1 & 2 & 0 & -72 \\ & 3 & 6 & 24 & 72 \\ \hline 1 & 2 & 8 & 24 \end{array}
$$

$$x^3 + 2x^2 + 8x + 24$$

9.

$$
2 \,) \, \begin{array}{rrrrrr} -3 & 0 & 0 & 0 & 7 & 8 \\ & -6 & -12 & -24 & -48 & -82 \\ \hline -3 & -6 & -12 & -24 & -41 & -74 \end{array}
$$

$$-3x^4 - 6x^3 - 12x^2 - 24x - 41 - \frac{74}{x - 2}$$

11.

$$
-\tfrac{1}{3} \,) \, \begin{array}{rrrr} 3 & -20 & 23 & 10 \\ & -1 & 7 & -10 \\ \hline 3 & -21 & 30 \end{array}
$$

$$3x^2 - 21x + 30$$

13.

$$
3 \,) \, \begin{array}{rrrr} 1 & 0 & 2 & -5 \\ & 3 & 9 & 33 \\ \hline 1 & 3 & 11 & \underline{28} = P(3) \end{array}
$$

$$P(3) = (3)^3 + 2(3) - 5 = 28$$

15.

$$
\begin{array}{r|rrrr}
\frac{1}{2} & 3 & 2 & -4 & 1 \\
  &   & \frac{3}{2} & \frac{7}{4} & -\frac{9}{8} \\
\hline
  & 3 & \frac{7}{2} & -\frac{9}{4} & \underline{\frac{1}{8}} = P\left(\frac{1}{2}\right)
\end{array}
$$

$$P\left(\tfrac{1}{2}\right) = 3\left(\tfrac{1}{2}\right)^3 + 2\left(\tfrac{1}{2}\right)^2 - 4\left(\tfrac{1}{2}\right) + 1 = -\tfrac{1}{8}$$

17.

$$
\begin{array}{r|rrrrr}
.1 & -6 & 0 & 5 & 0 & -2 \\
\hline
   & -6 & -.6 & 4.94 & .494 & \underline{-1.9506} = P(.1)
\end{array}
$$

$$P(.1) = -6(.1)^4 + 5(.1)^2 - 2 = -1.9506$$

19.

$$
\begin{array}{r|rrr}
3i & 1 & -2 & -1 \\
   &   & 3i & -6i-9 \\
\hline
   & 1 & -2+3i & \underline{-6i-10} = P(3i)
\end{array}
$$

$$P(3i) = (3i)^2 - 2(3i) - 1 = -6i - 10$$

21.

$$
\begin{array}{r|rrrr}
-2+i & 1 & 5 & 9 & 5 \\
     &   & -2+i & -7+i & -5 \\
\hline
     & 1 & 3+i & 2+i & \underline{0} = P(-2+i)
\end{array}
$$

$$P(-2+i) = (-2+i)^3 + 5(-2+i)^2 + 9(-2+i) + 5$$

23. $P(1) = (1)^3 - 2(1)^2 + 3k(1) - 3$

$$= 1 - 2 + 3k - 3 = 0, \quad k = \tfrac{4}{3}$$

25. $P(\sqrt{2}) = (\sqrt{2})^3 - 2k(\sqrt{2}) + 5 = 0$

$$-2\sqrt{2}k = -5 - 2\sqrt{2}$$

$$k = \frac{5 + 2\sqrt{2}}{2\sqrt{2}} = 1 + \frac{5}{2\sqrt{2}}$$

27. $P(x) = x^3 + 8x + 12$

$s_1 = 1(2) = 2$

$s_2 = 2 + 0 = 2$

$s_3 = 2(2) = 4$

$s_4 = 4 + 8 = 12$

$s_5 = 12(2) = 24$

$s_6 = 24 + 12 = 36$

29. $P(x) = 3x^7 - 2x^4 + 7x^2 - 5$

$s_1 = 3(-3) = -9$ $s_8 = 249 + 0$

$s_2 = -9 + 0 = -9$ $s_9 = 249(-3) = -747$

$s_3 = -9(-3) = 27$ $s_{10} = -747 + 7 = -740$

$s_4 = 27 + 0 = 27$ $s_{11} = -740(-3) = 2220$

$s_5 = 27(-3) = -81$ $s_{12} = 2220 + 0$

$s_6 = -81 + (-2) = -83$ $s_{13} = 2220(-3) = -6660$

$s_7 = (-83)(-3) = 249$ $s_{14} = -6660 - 5 = -6665 = P(-3)$

31. $s_1 = 2(2.4) = 4.8$

$s_2 = 4.8 + (-3) = 1.8$

$s_3 = (1.8)(2.4) = 4.32$

$s_4 = 4.32 + 0 = 4.32$

$s_5 = 4.32(2.4) = 10.368$

$s_6 = 10.368 + 5 = 15.368$

$s_7 = 15.368(2.4) = 36.8832$

$s_8 = 36.8832 + 25 = 61.8832$

1. $x^3 + 2x^2 - x - 2 = 0$

Possible ratio. roots: $\pm 1, 2$

$$
\begin{array}{r|rrrr}
1 & 1 & 2 & -1 & -2 \\
  &   & 1 & 3 & 2 \\
\hline
  & 1 & 3 & 2 & 0
\end{array}
$$

$$(x-1)(x^2 + 3x + 2) = 0$$
$$(x-1)(x+1)(x+2) = 0$$
$$x = 1, \, -1, \, -2$$

3. $y^3 + 2y^2 + y + 2 = 0$

Possible ratio. roots: $\pm 1, 2$

$$
\begin{array}{r|rrrr}
-2 & 1 & 2 & 1 & 2 \\
   &   & -2 & 0 & -2 \\
\hline
   & 1 & 0 & 1 &
\end{array}
$$

$$(y+2)(y^2 + 1) = 0$$
$$y = -2, \, \pm i$$

5. $2x^3 + x^2 - 18x - 20 = 0$

Possible ratio. roots: $\pm 1, 2, 4, 5, 10, \frac{1}{2}, \frac{5}{2}$

$$
\begin{array}{r|rrrr}
-\frac{5}{2} & 2 & 1 & -18 & 20 \\
             &   & -5 & 10 & 20 \\
\hline
             & 2 & -4 & -8 &
\end{array}
$$

$$(2x^2 - 4x - 8)\left(x+\tfrac{5}{2}\right) = 0$$
$$x = -\tfrac{5}{2}, \quad \frac{4 \pm \sqrt{16+64}}{4}$$
$$x = -\tfrac{5}{2}, \quad 1 \pm \sqrt{5}$$

7. $z^4 - 2z^3 - 7z^2 + 8z + 12 = 0$

Possible ratio. roots: $\pm 1, 2, 3, 4, 6, 12$

$$
\begin{array}{r|rrrrr}
-1 & 1 & -2 & -7 & 8 & 12 \\
   &   & -1 & 3 & 4 & -12 \\
\hline
   & 1 & -3 & -4 & 12 & 0 \\
\end{array}
$$

$(x+1)(x^3 - 3x^2 - 4x + 12) = 0$

$$
\begin{array}{r|rrrr}
2 & 1 & -3 & -4 & 12 \\
  &   & 2 & -2 & -12 \\
\hline
  & 1 & -1 & -6 & 0 \\
\end{array}
$$

$(x+1)(x-2)(x^2 - x - 6) = 0$

$(x+1)(x-2)(x-3)(x+2) = 0$

$x = -1, -2, 2, 3$

9. $3p^3 - 10p^2 + 12p - 3 = 0$

Possible ratio. roots: $\pm 1, 3, \frac{1}{3}$

$$
\begin{array}{r|rrrr}
\frac{1}{3} & 3 & -10 & 12 & -3 \\
  &   & 1 & -3 & 3 \\
\hline
  & 3 & -9 & 9 & 0 \\
\end{array}
$$

$\left(x - \frac{1}{3}\right)3(x^2 - 3x + 3) = 0$

$x = \frac{1}{3}, \quad \dfrac{3 \pm \sqrt{9-12}}{2}$

$x = \frac{1}{3}, \quad \dfrac{3 \pm i\sqrt{3}}{2}$

11. $x^4 - 6x^2 - 8x - 3 = 0$

Possible ratio. roots: $\pm 1, 3$

$$
\begin{array}{r|rrrrr}
-1 & 1 & 0 & -6 & -8 & -3 \\
   &   & -1 & 1 & 5 & 3 \\
\hline
   & 1 & -1 & -5 & -3 & 0 \\
\end{array}
$$

$$(x+1)(x^3 - x^2 - 5x - 3) = 0$$

$$
\begin{array}{r|rrrr}
-1 & 1 & -1 & -5 & -3 \\
   &   & -1 & 2 & 3 \\
\hline
   & 1 & -2 & -3 & 0 \\
\end{array}
$$

$$(x+1)^2(x^2 - 2x - 3) = 0$$
$$(x+1)^2(x+1)(x-3) = 0$$
$$(x+1)^3(x-3) = 0$$
$$x = -1, \ -1, \ -1, \ 3$$

13. $9w^3 - w + 2 = 0$

Possible ratio. roots: $\pm 1, 2, \frac{1}{9}, \frac{1}{3}, \frac{2}{9}, \frac{2}{3}$

$$
\begin{array}{r|rrrr}
-\frac{2}{3} & 9 & 0 & -1 & 2 \\
   &   & -6 & 4 & -2 \\
\hline
   & 9 & -6 & 3 & \\
\end{array}
$$

$$\left(x+\tfrac{2}{3}\right)3(3x^2 - 2x + 1) = 0$$

$$x = -\tfrac{2}{3}, \quad \frac{2 \pm \sqrt{4-12}}{6}$$

$$x = -\tfrac{2}{3}, \quad \frac{1 \pm i\sqrt{2}}{3}$$

15. $2x^3 - 3x^2 - 5x - 2 = 0$

Possible ratio. roots: $\pm 1, 2, \frac{1}{2}$

$$\frac{1}{2} \overline{\big)\quad 2 \quad -3 \quad 5 \quad -2\,}$$

	2	−3	5	−2
		1	−1	2
	2	−2	4	0

$\left(x - \frac{1}{2}\right)(x^2 - x + 2) = 0$

$$x = \frac{1}{2}, \quad \frac{1 \pm \sqrt{1-8}}{2}$$

$$x = \frac{1}{2}, \quad \frac{1 \pm i\sqrt{7}}{2}$$

17. $x^5 - x^3 - 8x^2 + 8 = 0$

Possible ratio. roots: $\pm 1, 2, 4, 8$

1	1	0	−1	−8	0	8
		1	1	0	−8	−8
	1	1	0	−8	−8	

$(x-1)(x^4 + x^3 - 8x - 8) = 0$

−1	1	1	0	−8	−8
		−1	0	0	8
	1	0	0	−8	

$(x-1)(x+1)(x^3 - 8) = 0$

$(x-1)(x+1)(x-2)(x^2 + 2x + 4) = 0$

$$x = 1, -1, 2, \quad \frac{-2 \pm \sqrt{4-16}}{2}$$

$$x = 1, -1, 2, -1 \pm i\sqrt{3}$$

19. $x^3 + 6x^2 + 2x - 3 = 0$

Possible ratio. roots: $\pm 1, 3$

$$
\begin{array}{r|rrrr}
-1 & 1 & 6 & 2 & -3 \\
& & -1 & -5 & 3 \\
\hline
& 1 & 5 & -3 &
\end{array}
$$

$$(x+1)(x^2 + 5x - 3) = 0$$

$$x = -1, \quad \frac{-5 \pm \sqrt{25+12}}{2}$$

$$x = -1, \quad \frac{-5 \pm \sqrt{37}}{2}$$

$$(x+1)\left(x+\frac{5}{2}+\frac{\sqrt{37}}{2}\right)\left(x+\frac{5}{2}-\frac{\sqrt{37}}{2}\right)$$

21. $4x^3 + 8x^2 + 9x + 3 = 0$

Possible ratio. roots: $\pm 1, 3, \frac{1}{4}, \frac{3}{4}, \frac{1}{2}$

$$
\begin{array}{r|rrrr}
-\frac{1}{2} & 4 & 8 & 9 & 3 \\
& & -2 & -3 & -3 \\
\hline
& 4 & 6 & 6 &
\end{array}
$$

$$\left(x-\frac{1}{2}\right)2(2x^2 + 3x + 3) = 0$$

$$x = -\frac{1}{2}, \quad \frac{-3 \pm i\sqrt{15}}{4}$$

$$2(2x+1)\left(x+\frac{3}{4}+\frac{i\sqrt{15}}{4}\right)\left(x+\frac{3}{4}-\frac{i\sqrt{15}}{4}\right)$$

23. $4x^4 - 4x^3 - 57x^2 + 81x = 0$

$x(4x^3 - 4x^2 - 57x + 81) = 0$

Possible ratio. roots: $\pm 1, 3, 9, 27, 81, \frac{1}{2}, \frac{3}{2}, \frac{9}{2}, \frac{27}{2}, \frac{81}{2}, \frac{1}{4}, \frac{3}{4}, \frac{9}{4}, \frac{27}{4}, \frac{81}{4}$

$$
\begin{array}{r|rrrr}
\frac{3}{2} & 4 & -4 & -57 & 81 \\
& & 6 & 3 & -81 \\
\hline
& 4 & 2 & -54 &
\end{array}
$$

$$x\left(x-\frac{3}{2}\right)2(2x^2 + x - 27) = 0$$

$$x = 0, \frac{3}{2}, \quad \frac{-1 \pm \sqrt{217}}{4}$$

$$2x(2x-3)\left(x+\frac{1}{4}+\frac{\sqrt{217}}{4}\right)\left(x+\frac{1}{4}-\frac{\sqrt{217}}{4}\right)$$

25. $y = x^4$

$y = 2x - x^3$

$$x^4 = 2x - x^3$$

$$x^4 + x^3 - 2x = 0$$

$$x(x^3 + x^2 - 2) = 0$$

Possible ratio. roots: $\pm 1, 2$

$$
\begin{array}{c|cccc}
1 & 1 & 1 & 0 & -2 \\
  &   & 1 & 2 & 2 \\
\hline
  & 1 & 2 & 2 &
\end{array}
$$

$$x(x-1)(x^2 + 2x + 2) = 0$$

$$x = 0, \quad 1, \quad -1 \pm i$$

$x = 0, \quad y = 0$ $x = -1 + i, \quad y = (-1 + i)^4 = -4$

$x = 1, \quad y = 1$ $x = -1 - i, \quad y = (-1 - i)^4 = -4$

27. $x^2 - xy = y + 1$

$xy + x = 4y^2$

$x^2 - 1 = y + xy$

$x^2 - 1 = y(1+x)$

$\frac{x^2 - 1}{1+x} = y$

$$y = x - 1, \quad x \neq -1$$

$$xy + x = 4y^2$$

$$x(x-1) + x = 4(x-1)^2$$

$$x^2 - x + x = 4x^2 - 8x + 4$$

$$3x^2 - 8x + 4 = 0$$

$$(3x-2)(x-2) = 0$$

$$x = 2, \quad \tfrac{2}{3}$$

$x = 2, \quad y = 2 - 1 = 1$ $(2, 1)$

$x = \tfrac{2}{3}, \quad y = \tfrac{2}{3} - 1 = -\tfrac{1}{3}$ $\left(\tfrac{2}{3}, -\tfrac{1}{3}\right)$

Consider $x = -1$

$x = -1, \quad -y - 1 = 4y^2$

$$4y^2 + y + 1 = 0$$

$$y = \frac{-1 \pm \sqrt{15i}}{8}$$

1. $x^2 + 3x + 2 = 0$ no positive

$P(-x) = -x^3 - 3x + 2$ 1 negative, 2 complex

3. $x^4 + 13x^2 - 7 = 0$ 1 positive

$P(-x) = x^4 + 13x^2 - 7$ 1 negative

So 2 complex

5. $2x^3 - 7x^2 - 3x - 1 = 0$ 1 positive

$P(-x) = -2x^3 - 7x^2 + 3x - 1$ 2 or 0 negative

So 1 positive, 2 complex   or   1 positive, 2 negative

7. $-2x^5 + 8x^3 - 7x + 15 = 0$ 3 or 1 positive

$P(-x) = 2x^5 - 8x^3 + 7x + 15$ 2 or 0 negative

3 positive, 2 complex   or   3 positive, 2 negative   or
1 positive, 4 complex   or   1 positive, 2 negative, 2 complex

9. $3x^6 + x^4 - 8x^2 + 16 = 0$ 2 or 0 positive

$P(-x) = 3x^6 + x^4 - 8x^2 + 16$ 2 or 0 negative

2 positive, 4 complex   or   2 positive, 2 negative, 2 complex   or
2 negative, 4 complex   or   6 complex

11. $x^3 + 2x^2 + 8x - 3 = 0$

$P(0) = -3,$ $P(1) = 8$

$c = \dfrac{0(8) - 1(-3)}{8 + 3} = \dfrac{3}{11} = .27$

13. $x^3 - 3x + 1 = 0$

$P(1) = -1,$ $P(2) = 3$

$c = \dfrac{1(3) - 2(-1)}{3 + 1} = \dfrac{5}{4} = 1.25$

15. $2x^3 - 3x^2 - 12x + 6 = 0$

$P(3) = 2(27) - 3(9) - 12(3) + 6 = -3$

$P(4) = 2(64) - 3(16) - 12(4) + 6 = 38$

$c = \dfrac{3(38) - 4(-3)}{38 + 3} = \dfrac{126}{41} = 3.07$

17. $x^4 - x^3 - 5x^2 + 3x + 2 = 0$

$P(-1) = 1 + 1 - 5 - 3 + 2 = -4$

$P(0) = 2$

$c = \dfrac{-1(2) - 0(-4)}{2 + 4} = -\dfrac{2}{6} = -.33$

19. $x^4 - x^3 - 2x^2 - x - 3 = 0$

$P(2) = 16 - 8 - 8 - 2 - 3 = -5$

$P(3) = 81 - 27 - 2(9) - 3 - 3 = 24$

$c = \dfrac{2(24) - 3(-5)}{24 + 5} = \dfrac{63}{29} = 2.17$

21. $2.1x^3 + 3.4x^2 - 1.8x + 5.3 = 0$

$P(-3) = 2.1(-27) + 3.4(9) - 1.8(-3) + 5.3 = -15.4$

$P(-2) = 2.1(-8) + 3.4(-4) - 1.8(-2) + 5.3 = 15.7$

$c = \dfrac{-3(5.7) + 2(-15.4)}{5.7 + 15.4} = -\dfrac{47.9}{21.1} = -2.27$

Section 8.6   Decomposition of Rational Functions into Partial Fractions

1. $\dfrac{x + 1}{(x+5)(x-1)} = \dfrac{A}{x + 5} + \dfrac{B}{x - 1}$

$x + 1 = A(x-1) + B(x+5)$

$x + 1 = Ax - A + Bx + 5B$

$x + 1 = (A+B)x + (-A + 5B)$

$\begin{aligned} A + B &= 1 \\ -A + 5B &= 1 \end{aligned}$

$6B = 2, \quad B = \tfrac{1}{3}, \quad A = \tfrac{2}{3}$

3. $\dfrac{1}{x(x+1)} = \dfrac{A}{x} + \dfrac{B}{x+1}$

$1 = A(x+1) + Bx$

$1 = (A+B)x + A$

$A = 1, \quad A + B = 0, \quad B = -1$

5. $\dfrac{2x^2 + 5x + 5}{x(x+5)(x+2)} = \dfrac{A}{x} + \dfrac{B}{x+5} + \dfrac{C}{x+2}$

$2x^2 + 5x + 5 = A(x+5)(x+2) + Bx(x+2) + Cx(x+5)$

$2x^2 + 5x + 5 = Ax^2 + 7Ax + 10A + Bx^2 + 2Bx + Cx^2 + 5Cx$

$2x^2 + 5x + 5 = (A+B+C)x^2 + (7A+2B+5C)x + 10A$

$A + B + C = 2$

$7A + 2B + 5C = 5$

$10A = 5, \quad A = \tfrac{1}{2}$

$B + \quad C = \quad \tfrac{3}{2}$

$2B + 5C = \quad \tfrac{3}{2}$

$\rule{3cm}{0.4pt}$

$-2B - 2C = -3$

$2B + 5C = \quad \tfrac{3}{2}$

$\rule{3cm}{0.4pt}$

$3C = -\tfrac{3}{2}, \quad C = -\tfrac{1}{2}, \quad B = 2$

7. $\dfrac{x+3}{x^2 + 6x + 5} = \dfrac{x+3}{(x+5)(x+1)} = \dfrac{A}{x+5} + \dfrac{B}{x+1}$

$x + 3 = A(x+1) + B(x+5)$

$x + 3 = Ax + A + Bx + 5B$

$x + 3 = (A+B)x + A + 5B$

$A + \quad B = \quad 1$

$A + 5B = \quad 3$

$\rule{3cm}{0.4pt}$

$-4B = -2, \quad B = \tfrac{1}{2}, \quad A = \tfrac{1}{2}$

9. $\dfrac{x^2 + 3x - 6}{x(x-1)^2} = \dfrac{A}{x} + \dfrac{B}{x-1} + \dfrac{C}{(x-1)^2}$

$x^2 + 3x - 6 = A(x-1)^2 + Bx(x-1) + Cx$

$x^2 + 3x - 6 = Ax^2 - 2Ax + A + Bx^2 - Bx + Cx$

$x^2 + 3x - 6 = (A+B)x^2 + (-2A - B + C)x + A$

$A + B = 1$

$-2A - B + C = 3$

$A = -6, \quad B = 7$

$C = 3 + 2A + B = -2$

11. $\dfrac{1}{x(x^2 + 2x + 1)} = \dfrac{1}{x(x+1)^2} = \dfrac{A}{x} + \dfrac{B}{x+1} + \dfrac{C}{(x+1)^2}$

$1 = A(x+1)^2 + Bx(x+1) + Cx$

$1 = Ax^2 + 2Ax + A + Bx^2 + Bx + Cx$

$1 = (A+B)x^2 + (2A+B+C)x + A$

$A + B = 0$

$2A + B + C = 0$

$A = 1, \quad B = -1$

$C = -2A - B = -1$

13. $\dfrac{x^2 - 3x + 6}{(x-1)(x+1)(3-2x)} = \dfrac{A}{x-1} + \dfrac{B}{x+1} + \dfrac{C}{3-2x}$

$x^2 - 3x + 6 = A(x+1)(3-2x) + B(x-1)(3-2x) + C(x-1)(x+1)$

$x^2 - 3x + 6 = -2Ax^2 + Ax + 3A - 2Bx^2 + 5Bx - 3B + Cx^2 - C$

$x^2 - 3x + 6 = (-2A-2B+C)x^2 + (A+5B)x + (3A-3B-C)$

$-2A - 2B + C = 1$

$A + 5B = -3$

$\underline{3A - 3B - C = 6}$

$-2A - 2B + C = \quad 1$

$\underline{3A - 3B - C = \quad 6}$

$A - 5B \quad = \quad 7$

$\underline{A + 5B \quad = -3}$

$2A = 4, \quad A = 2$

$B = \dfrac{-3 - A}{5} = -\dfrac{5}{5} = -1, \quad C = 1 + 2A + 2B = 3$

15. $\dfrac{2x-1}{(x-1)^2} = \dfrac{A}{x-1} + \dfrac{B}{(x+1)^2}$

$2x+1 = A(x-1) + B$

$2x-1 = Ax - A + B$

$A = 2, \quad -A + B = -1$

$\qquad\qquad B = -1 + A = 1$

17. $\dfrac{x^2 + x + 2}{x(x^2 + 2)} = \dfrac{A}{x} + \dfrac{Bx + C}{x^2 + 2}$

$x^2 + x + 2 = A(x^2+2) + (Bx+C)x$

$x^2 + x + 2 = Ax^2 + 2A + Bx^2 + Cx$

$x^2 + x + 2 = (A+B)x^2 + Cx + 2A$

$\qquad A + B = 1$

$\qquad\qquad C = 1$

$\qquad 2A = 2, \quad A = 1, \quad B = 0$

19. $\dfrac{x^2 + 1}{(2x+3)(x^2+2x+4)} = \dfrac{A}{2x + 3} + \dfrac{Bx + C}{x^2 + 2x + 4}$

$x^2 + 1 = A(x^2+2x+4) + (Bx+C)(2x+3)$

$x^2 + 1 = Ax^2 + 2Ax + 4A + 2Bx^2 + 2Cx + 3Bx + 3C$

$x^2 + 1 = (A+2B)x^2 + (2A+3B+2C)x + (4A+3C)$

$\qquad A + 2B = 1$

$2A + 3B + 2C = 0$

$\qquad 4A + 3C = 1$

$A = \dfrac{\begin{vmatrix} 1 & 2 & 0 \\ 0 & 3 & 2 \\ 1 & 0 & 3 \end{vmatrix}}{\begin{vmatrix} 1 & 2 & 0 \\ 2 & 3 & 2 \\ 4 & 0 & 3 \end{vmatrix}} = \dfrac{13}{13} = 1$

$B = \dfrac{\begin{vmatrix} 1 & 1 & 0 \\ 2 & 0 & 2 \\ 4 & 1 & 3 \end{vmatrix}}{13} = \dfrac{0}{13} = 0$

$C = \dfrac{\begin{vmatrix} 1 & 2 & 1 \\ 2 & 3 & 0 \\ 4 & 0 & 1 \end{vmatrix}}{13} = -\dfrac{13}{13} = -1$

21. $\dfrac{x^3 - 3x^2 + 7x - 5}{(x^2 + 2x + 3)(x^2 + 5)} = \dfrac{Ax + B}{x^2 + 2x + 3} + \dfrac{Cx + D}{x^2 + 5}$

$x^3 - 3x^2 + 7x - 5 = (Ax+B)(x^2+5) + (Cx+D)(x^2+2x+3)$

$x^3 - 3x^2 + 7x - 5 = Ax^3 + 5Ax + Bx^2 + 5B + Cx^3 + 2Cx^2 + 3Cx + Dx^2 + 2Dx + 3D$

$x^3 - 3x^2 + 7x - 5 = (A+C)x^3 + (B+2C+D)x^2 + (5A+3C+2D)x + (5B+3D)$

$$A + C = 1$$
$$B + 2C + D = -3$$
$$5A + 3C + 2D = 7$$
$$5A + 3D = -5$$

$$A = \dfrac{\begin{vmatrix} 1 & 0 & 1 & 0 \\ -3 & 1 & 2 & 1 \\ 7 & 0 & 3 & 2 \\ -5 & 5 & 0 & 3 \end{vmatrix}}{\begin{vmatrix} 1 & 0 & 1 & 0 \\ 0 & 1 & 2 & 1 \\ 5 & 0 & 3 & 2 \\ 0 & 5 & 0 & 3 \end{vmatrix}} = \dfrac{1\begin{vmatrix} 1 & 2 & 1 \\ 0 & 3 & 2 \\ 5 & 0 & 3 \end{vmatrix} + 1\begin{vmatrix} -3 & 1 & 1 \\ 7 & 0 & 2 \\ -5 & 5 & 3 \end{vmatrix}}{1\begin{vmatrix} 1 & 2 & 1 \\ 0 & 3 & 2 \\ 5 & 0 & 3 \end{vmatrix} + 1\begin{vmatrix} 0 & 1 & 1 \\ 5 & 0 & 2 \\ 0 & 5 & 3 \end{vmatrix}} = \dfrac{48}{24} = 2$$

$$B = \dfrac{\begin{vmatrix} 1 & 1 & 1 & 0 \\ 0 & -3 & 2 & 1 \\ 5 & 7 & 3 & 2 \\ 0 & -5 & 0 & 3 \end{vmatrix}}{24} = \dfrac{-24}{24} = -1$$

$$C = \dfrac{\begin{vmatrix} 1 & 0 & 1 & 0 \\ 0 & 1 & -3 & 1 \\ 5 & 0 & 7 & 2 \\ 0 & 5 & -5 & 3 \end{vmatrix}}{24} = \dfrac{-24}{24} = -1$$

$$D = \dfrac{\begin{vmatrix} 1 & 0 & 1 & 1 \\ 0 & 1 & 2 & -3 \\ 5 & 0 & 3 & 7 \\ 0 & 5 & 0 & -5 \end{vmatrix}}{24} = \dfrac{0}{24} = 0$$

1.

$$1 \overline{)\ \begin{array}{ccc} 1 & -3 & 4 \\ & 1 & -2 \end{array}}$$
$$\begin{array}{ccc} 1 & -2 & 2 \end{array}$$

$x^2 - 3x + 4 \div (x-1) = x - 2 + \dfrac{2}{x-1}$

$P(1) = 1 - 3 + 4 = 2$

3.

$$2 \overline{)\ \begin{array}{ccc} 3 & -1 & 4 \\ & 6 & 10 \end{array}}$$
$$\begin{array}{ccc} 3 & 5 & 14 \end{array}$$

$3x^2 - x + 4 \div (x-2) = 3x + 5, \quad R = 14$

$P(2) = 3(4) - 2 + 4 = 14$

5.

$$\tfrac{1}{2} \overline{)\ \begin{array}{cccc} 2 & 0 & -1 & 1 \\ & 1 & \tfrac{1}{2} & \tfrac{1}{4} \end{array}}$$
$$\begin{array}{cccc} 2 & 1 & -\tfrac{1}{2} & \tfrac{3}{4} \end{array}$$

$2x^3 - x + 1 \div (2x-1) = 2x^2 + x - \dfrac{1}{2}, \quad R = \dfrac{3}{4}$

$P\!\left(\dfrac{1}{2}\right) = 2\!\left(\dfrac{1}{8}\right) - \dfrac{1}{2} + 1 = \dfrac{3}{4}$

7.

$$-1 \overline{)\ \begin{array}{ccccc} 3 & 1 & 0 & 1 & -1 & -2 \\ & -3 & 2 & -2 & 1 & 0 \end{array}}$$
$$\begin{array}{cccccc} 3 & -2 & 2 & -1 & 0 & -2 \end{array}$$

$3x^5 + x^4 + x^2 - x - 2 \div (x+1) = 3x^4 - 2x^3 + 2x^2 - x, \quad R = -2$

$P(-1) = -3 + 1 + 1 - 2 = -2$

9.

$$\begin{array}{r} \frac{2}{3}x^3 \ + \ \frac{2}{9}x^2 \ + \ \frac{2}{27}x \ + \ \frac{29}{81} \\ 3x - 1 \overline{\smash{\big)}\ 2x^4 \qquad\qquad\qquad + \ x \ - \ 1} \end{array}$$

$$\begin{array}{r} 2x^4 \ - \ \frac{2x^3}{3} \\ \hline \frac{2}{3}x^3 \\ \frac{2}{3}x^3 \ - \ \frac{2x^2}{9} \\ \hline \frac{2x^2}{9} \ + \ x \\ \frac{2x^2}{9} \ - \ \frac{2}{27}x \\ \hline \frac{29}{27}x \ - \ 1 \\ \frac{29}{27}x \ - \ \frac{29}{81} \\ \hline - \frac{52}{81} = R \end{array}$$

$$P\left(\frac{1}{3}\right) = 2\left(\frac{1}{81}\right) + \frac{1}{3} - 1 = -\frac{52}{81}$$

11. $P(3) = 3^2 - 3(3) + 2 = +2,$ No factor

13. $P(2) = 2^4 - 16 = 0,$ Yes

15. $P\left(-\frac{1}{2}\right) = 2\left(-\frac{1}{2}\right)^3 - \left(-\frac{1}{2}\right)^2 - \frac{1}{2} - 1 = -2,$ No factor

17. $P(x) = a(x-2)(x-2-i)(x-2+i)(x-i)(x+i)$

$P(0) = 1 = a(-2)(-2-i)(-2+i)(-i)(i),$ $a = -\frac{1}{10}$

$P(x) = -\frac{1}{10}(x-2)(x^2-4x+5)(x^2+1)$

19. $P(x) = a(x-i)(x+i)(x-2i)(x+2i)$

$P(1) = 20 = a(1-i)(1+i)(1-2i)(1+2i),$ $a = 2$

$P(x) = 2(x^2+1)(x^2+4)$

21. $P(x) = a(x+2)(x+1)(x-1)(x-3)(x-4)$

$P(0) = -25 = a(2)(1)(-1)(-3)(-4),$ $a = \frac{25}{24}$

$P(x) = \frac{25}{24}(x+2)(x+1)(x-1)(x-3)(x-4)$

23. $x^4 + x^3 - x^2 + x - 2 = 0$

Possible ratio. roots: $\pm 1, 2$

$$
\begin{array}{r|rrrrr}
1 & 1 & 1 & -1 & 1 & -2 \\
 & & 1 & 2 & 1 & 2 \\
\hline
-2 & 1 & 2 & 1 & 2 \\
 & & -2 & 0 & -2 \\
\hline
 & 1 & 0 & 1 \\
\end{array}
$$

$(x-1)(x+2)(x^2+1) = 0$

$x = 1, -2, \pm i$

25. $x^4 + x^3 - 3x^2 - x + 2 = 0$

Possible ratio. roots: $\pm 1, 2$

$$
\begin{array}{r|rrrrr}
1 & 1 & 1 & -3 & -1 & 2 \\
 & & 1 & 2 & -1 & -2 \\
\hline
1 & 1 & 2 & -1 & -2 \\
 & & 1 & 3 & 2 \\
\hline
 & 1 & 3 & 2 \\
\end{array}
$$

$(x-1)^2(x^2+3x+2) = 0$

$(x-1)^2(x+2)(x+1) = 0$

$x = +1, +1, -2, -1$

27. $x^3 - 7x^2 + 5x + 1 = 0$

Possible ratio. roots: ± 1

$$
\begin{array}{r|rrrr}
1 & 1 & -7 & 5 & 1 \\
 & & 1 & -6 & -1 \\
\hline
 & 1 & -6 & 1 \\
\end{array}
$$

$(x-1)(x^2-6x-1) = 0$

$x = 1, \quad \dfrac{6\pm\sqrt{40}}{2} = 3\pm\sqrt{10}$

29. $2x^3 - 3x^2 + 2x + 2 = 0$

Possible ratio. roots: $\pm 1, 2, \frac{1}{2}$

$$
-\frac{1}{2} \overline{) \begin{array}{rrrr} 2 & -3 & 2 & 2 \\ & -1 & 2 & 0 \\ \hline 2 & -4 & 4 & \end{array}}
$$

$\left(x+\frac{1}{2}\right)2(x^2-2x+2) = 0$

$x = -\frac{1}{2}, \quad \frac{2 \pm \sqrt{-4}}{2} = -\frac{1}{2}, \quad 1 \pm i$

31. $P(x) = x^4 - x - 3, \quad 1$ positive

$P(-x) = x^4 + x - 3, 1$ negative,

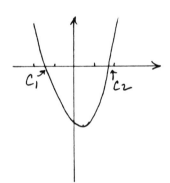

$P(2) = \left. \begin{array}{r} 11 \\ \end{array} \right]$

$P(1) = \left. \begin{array}{r} -3 \\ \end{array} \right]$

$c_2 = \dfrac{1(11) - (2)(-3)}{11 + 3} = 1.21$

$P(0) = -3$

$P(-1) = \left. \begin{array}{r} -1 \\ \end{array} \right]$

$P(-2) = \left. \begin{array}{r} 15 \\ \end{array} \right]$

$c_1 = \dfrac{-2(-1) - (-1)(15)}{-1 - 15} = -1.06$

33. $P(x) = x^5 - 1.1,$ 1 positive

$P(-x) = -x^5 - 1.1,$ 0 negative

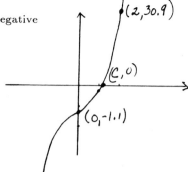

$P(0) = -1.1$

$P(1) = -.1$

$P(2) = 30.9$

$$c = \frac{1(30.9) - 2(-.1)}{30.9 + .1} = 1.003$$

35. $P(x) = 2.3x^3 - 7x + 1.8,$ 2 positive

$P(-x) = -2.3x^3 + 7x + 1.8,$ 1 negative

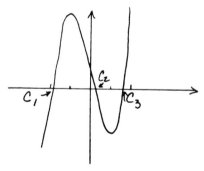

$P(-2) = -2.6$

$P(-1) = 6.5$

$$c_1 = \frac{-2(6.5) - (-1)(-2.6)}{6.5 + 2.6} = -1.71$$

$P(0) = 1.8$

$P(1) = -2.9$

$$c_2 = \frac{0(-2.9) - 1(1.8)}{-2.9 - 1.8} = .38$$

$P(1) = -2.9$

$P(2) = 6.2$

$$c_3 = \frac{1(6.2) - 2(2.9)}{6.2 + 2.9} = 1.32$$

37. $\dfrac{8x + 7}{(x-1)(x^2+4)} = \dfrac{A}{x-1} + \dfrac{Bx+C}{x^2+4}$

$8x + 7 = A(x^2+4) + (Bx+C)(x-1)$

$8x + 7 = Ax^2 + 4A + Bx^2 - Bx + Cx - C$

$8x + 7 = (A+B)x^2 + (-B+C)x + (4A-C)$

$A + B = 0, \quad A = -B$
$-B + C = 8$
$4A - C = 7$

$\quad -B + C = 8$
$\quad \underline{-4B - C = 7}$
$\quad -5B \quad = 15, \quad B = -3, \quad A = 3, \quad C = 8+B = 5$

39. $\dfrac{x^2 + 3x - 9}{x^3 - x^2 + 4x - 4} = \dfrac{x^2 + 3x - 9}{x^2(x-1) + 4(x-1)} = \dfrac{x^2 + 3x - 9}{(x^2+4)(x-1)} = \dfrac{A}{x-1} + \dfrac{Bx+C}{x^2+4}$

$x^2 + 3x - 9 = A(x^2+4) + (Bx+C)(x-1)$

$x^2 + 3x - 9 = Ax^2 + 4A + Bx^2 - Bx + Cx - C$

$x^2 + 3x - 9 = (A+B)x^2 + (C-B)x + (4A-C)$

$A + B = 1$
$C - B = 3$
$4A - C = -9$

$\quad A + B = 1$
$\quad \underline{C - B = 3}$
$\quad A + C = 4$
$\quad \underline{4A - C = -9}$
$\quad 5A \quad = -5, \quad A = -1, \quad B = 2, \quad C = 5$

41. $P(x) = x - \dfrac{1}{3}x^3$

$P(.1) = .1 - \dfrac{1}{3}(.1)^3 = .099667, \quad \sin .1 = .099833$

$P(.2) = .2 - \dfrac{1}{3}(.2)^3 = .197333, \quad \sin .2 = .198669$

$P(.5) = .5 - \dfrac{1}{3}(.5)^3 = .458333, \quad \sin .5 = .479426$

43. $\frac{1}{2} = \frac{1}{R_2 - 1} + \frac{1}{R_2} + \frac{1}{R_2 + 1}$

$(R_2 - 1)(R_2)(R_2 + 1) = 2R_2(R_2 + 1) + (R_2{}^2 - 1)2 + 2R_2(R_2 - 1)$

$R_2{}^3 - R_2 = 2R_2{}^2 + \cancel{2R_2} + 2R_2{}^2 - 2 + 2R_2{}^2 - \cancel{2R_2}$

$R_2{}^3 - 6R_2{}^2 - R_2 + 2 = 0$ 2 positive roots

$-R_2{}^3 - 6R_2{}^2 + R_2 + 2 = 0$ 1 negative root

Because $R_1 = R_2 - 1$ and $R_1 > 0, \quad R_2 > 1$

$P(6) = -4$

$P(7) = 42$

$c = \frac{6(42) - (7)(-4)}{42 + 4} = \frac{280}{46} = 6.09$

45. $v_1 = \frac{4\pi r^3}{3}, \quad v_2 = \frac{4\pi(r+1)^3}{3}$

$v_2 - v_1 = \frac{4\pi(r+1)^3}{3} - \frac{4\pi r^3}{3} = 15$

$\frac{4\pi(r^3 + 3r^2 + 3r + 1)}{3} - \frac{4\pi r^3}{3} - 15 = 0$

$4\pi\left(r^2 + r + \frac{1}{3}\right) - 15 = 0$

$4\pi\left(r^2 + r + \frac{1}{3} - \frac{15}{4\pi}\right) = 0$

$r = \frac{-1 \pm \sqrt{1 - 4\left(\frac{1}{3} - \frac{15}{4\pi}\right)}}{2}$,   negative root cannot be a radius

$r = \frac{-1 + \sqrt{1 + 3.441}}{2} = .553$

Chapter 9 Sequences, Probability and Mathematical Induction

Section 9.1   Sequences in General

1. $a_n = n + 1$

   $a_1 = 1 + 1 = 2, \quad a_2 = 2 + 1 = 3, \quad a_3 = 3 + 1 = 4$

   $a_4 = 4 + 1 = 5, \quad a_5 = 5 + 1 = 6$

3. $a_n = 1 + (-1)^n$

   $a_1 = 1 - (-1)^1 = 2, \quad a_2 = 1 - (-1)^2 = 0, \quad a_3 = 1 - (-1)^3 = 2,$

   $a_4 = 1 - (-1)^4 = 0, \quad a_5 = 1 - (-1)^5 = 2$

5. $a_n = \cos n\pi$

   $a_1 = \cos \pi = -1, \quad a_2 = \cos 2\pi = 1, \quad a_3 = \cos 3\pi = -1,$

   $a_4 = \cos 4\pi = 1, \quad a_5 = \cos 5\pi = -1$

7. $a_n = \dfrac{n + 2}{n + 1}$

   $a_1 = \dfrac{1 + 2}{1 + 1} = \dfrac{3}{2}, \quad a_2 = \dfrac{2 + 2}{2 + 1} = \dfrac{4}{3}, \quad a_3 = \dfrac{3 + 2}{3 + 1} = \dfrac{5}{4},$

   $a_4 = \dfrac{4 + 2}{4 + 1} = \dfrac{6}{5}, \quad a_5 = \dfrac{5 + 2}{5 + 1} = \dfrac{7}{6}$

9. $a_n = n^2$

   $a_1 = 1^2 = 1, \quad a_2 = 2^2 = 4, \quad a_3 = 3^2 = 9, \quad a_4 = 4^2 = 16 \quad a_5 = 5^2 = 25$

11. $a_n = 2n + 1$

   $a_1 = 2(1) + 1 = 3, \quad a_2 = 2(2) + 1 = 5, \quad a_3 = 2(3) + 1 = 7,$

   $a_4 = 2(4) + 1 = 9, \quad a_5 = 2(5) + 1 = 11$

13. $a_n = \dfrac{1}{n^2}$

   $a_1 = \dfrac{1}{1} = 1, \quad a_2 = \dfrac{1}{2^2} = \dfrac{1}{4}, \quad a_3 = \dfrac{1}{3^2} = \dfrac{1}{9}, \quad a_4 = \dfrac{1}{4^2} = \dfrac{1}{16}, \quad a_5 = \dfrac{1}{5^2} = \dfrac{1}{25}$

15. $a_n = 5\left(\frac{1}{2}\right)^n$

$a_1 = 5\left(\frac{1}{2}\right)^1 = \frac{5}{2}, \quad a_2 = 5\left(\frac{1}{2}\right)^2 = \frac{5}{4}, \quad a_3 = 5\left(\frac{1}{2}\right)^3 = \frac{5}{8},$

$a_4 = 5\left(\frac{1}{2}\right)^4 = \frac{5}{16}, \quad a_5 = 5\left(\frac{1}{2}\right)^5 = \frac{5}{32}$

27. 1st year: $4000 - 4000(.25) = 4000(.75)$

    2nd year: $4000(.75)^2$

    3rd year: $4000(.75)^3$

    4th year: $4000(.75)^4$

Section 9.2   Sequences of Partial Sums

1. $\displaystyle\sum_{m=1}^{5} \frac{1}{m+2} = \frac{1}{3} + \frac{1}{4} + \frac{1}{5} + \frac{1}{6} + \frac{1}{7} = 1.093$

3. $\displaystyle\sum_{n=0}^{5} n^2 = 0 + 1^2 + 2^2 + 3^2 + 4^2 + 5^2 = 55$

5. $\displaystyle\sum_{j=1}^{5} 3j = 3\sum_{j=1}^{5} = 3(1+2+3+4+5) = 45$

7. $\displaystyle\sum_{k=1}^{5} (1+k) = \sum_{k=1}^{5} 1 + \sum_{k=1}^{5} = 5 + (1+2+3+4+5) = 20$

9. $\displaystyle\sum_{n=-1}^{2} n^2 = (-2)^2 + (-1)^2 + 0 + 1^2 + 2^2 = 10$

19. $\displaystyle\sum_{i=1}^{n} (x_i + d) = (x_1 + d) + (x_2 + d) + \ldots + (x_n + d)$

                $= x_1 + x_2 + \ldots + x_n + nd$

                $= \displaystyle\sum_{i=1}^{n} x_i + nd, \quad$ True

21. $\displaystyle\sum_{k=2}^{5} a_{k-2} = a_0 + a_1 + a_2 + a_3 = \sum_{k=0}^{3} a_k, \quad$ True

23. $\displaystyle\sum_{n=1}^{3} x^n = x^1 + x^2 + x^3$

$\displaystyle\sum_{t=2}^{4} x^{t-1} = x^1 + x^2 + x^3,$ True

27. $S_n = (-1)^n$

$S_1 = a_1 = -1$

$S_2 = S_1 + a_2 = 1$

     $-1 + a_2 = 1, \quad a_2 = 2$

$S_3 = S_2 + a_3 = -1$

     $1 + a_3 = -1, \quad a_3 = -2$

$S_4 = S_3 + a_4 = 1$

     $-1 + a_4 = 1, \quad a_4 = 2$

So $a_1 = -1, \quad a_n = 2(-1)^n$

Section 9.3   Arithmetic Sequences

1. $2-1=1, 3-2=1, d=1$

3. $4-2=2, 6-4=2, d=2$

5. $6-3=3, 9-6=3, d=3$

7. $\frac{1}{2}-1=-\frac{1}{2}, \frac{1}{4}-\frac{1}{2}=-\frac{1}{4}$, No

9. $\frac{1}{3}-\frac{1}{2}=-\frac{1}{6}, \frac{1}{4}-\frac{1}{3}=-\frac{1}{12}$, No

11. $a_1 = 2, d = 3$

    $2, 5, 8, 11\ldots$

    $a_n = 2 + (n-1)3$

      $= 3n - 1$

13. $S_n = a_1 n + \dfrac{n(n-1)d}{2}$

    $35 = 2(7) + \dfrac{7(6)d}{2}, \; d = 1$

15. $a_1 = a_2 - d$

    $a_1 = 5 - (-4) = 9$

    $S_{10} = 9(10) + \dfrac{10(9)(-4)}{2}$

      $= -90$

17. $\displaystyle\sum_{k=1}^{8} (2k-3)$ *difference*

    $a_1 = -1, \; a_2 = 1, \; d = 1 - (-1) = 2$

    $S_8 = (-1)(8) + \dfrac{(8)(7)(2)}{2} = 48$

19. $\displaystyle\sum_{k=0}^{4}(7k-4)$

$a_0 = -4$, $a_1 = 3$, $d = 3 - (-4) = 7$

$S_5 = -4(5) + \dfrac{(5)(4)(7)}{2} = 50$

a₀ = 5 places (handwritten: $a_0 = 5\ places$)

21. $\displaystyle\sum_{k=0}^{4}7(k-4)$

$a_0 = -28$, $a_1 = -21$,

$d = -21 - (-28) = 7$

$S_5 = -28(5) + \dfrac{(5)(4)(7)}{2} = -70$

(handwritten: a_1)

(handwritten top right: a_0 go 5, a_0 for a_1)

23. $\displaystyle\sum_{i=1}^{n}(2n-1)$

$a_1 = 1$, $d = 2$

$S_n = 1(n) + \dfrac{(n)(n-1)2}{2}$

$\quad = n + n(n-1) = n^2$

25. Let $a_1 = 1$, $a_2 = 4$, $a_3 = 9$

$d = a_2 - a_1 = 4-1 = 3$

$S_3 = 1(3) + \dfrac{3(2)3}{2} = 12$

or

$d = a_3 - a_2 = 9-4 = 5$

$S_3 = 1(3) = \dfrac{3(2)5}{2} = 18$

Actual $S_3 = 1 + 4 + 9 = 14$

27. $f(x) = x + 2$

$f(1) = 1+2 = 3$, $f(2) = 2+2 = 4$, $f(3) = 3+2 = 5$

$f(4) = 4+2 = 6$, $f(5) = 5+2 = 7$

29. $f(x) = -x+2$

$f(1) = -1+2 = 1$, $f(2) = -2+2 = 0$, $f(3) = -3+2 = -1$

$f(4) = -4+2 = -2$, $f(5) = -5+2 = -3$

31. $f(x) = 2x+5$

$f(1) = 2(1)+5 = 7$, $f(2) = 2(2)+5 = 9$, $f(3) = 2(3)+5 = 11$

$f(4) = 2(4)+5 = 13$, $f(5) = 2(5)+5 = 15$

33. $a_1 = 3$, $d = 6-3 = 3$

$f(x) = 3 + (x-1)3 = 3x$

35. $a_1 = 4$, $d = 3$

$f(x) = 4 + (x-1)3 = 3x+1$

37. $a_1 = 5$, $a_8 = 26$

$a_n = a_1 + (n-1)d$

$a_8 = 26 = 5+7d$, $d = 3$

$f(x) = 5 + (x-1)3 = 3x+2$

1. $r = \frac{a_n}{a_n-1} = \frac{12}{4} = 3$

$r = \frac{36}{12} = 3$,   yes, $a_n = 4(3)^{n-1}$

3. $r = \frac{6}{3} = 3$

$r = \frac{9}{6} = \frac{3}{2}$,   No

5. $r = \dfrac{-\frac{2}{3}}{\frac{1}{3}} = -2$

$r = \dfrac{\frac{4}{3}}{-\frac{2}{3}} = -2$,   yes, $a_n = \left(\frac{1}{3}\right)(-2)^{n-1}$

7. $r = \dfrac{\frac{1}{4}}{\frac{1}{2}} = \frac{1}{2}$

$r = \dfrac{\frac{1}{8}}{\frac{1}{4}} = \frac{1}{2}$,   yes, $a_n = \frac{1}{2}\left(\frac{1}{2}\right)^{n-1} = \left(\frac{1}{2}\right)^n$

9. $a_1 = 3$, $r = 5$

$a_4 = 3(5)^3 = 375$

$s_4 = \frac{1-5^4}{1-5}(3) = \frac{-624}{-4}(3) = 468$

11. $a_1 = 5$, $a_5 = 80$

$a_5 = 80 = 5r^4$

$r = \pm 2$

$r = 2$, $S_5 = \frac{(1-2^5)}{1-2}(5) = 155$

$r = -2$, $S_5 = \dfrac{\left(1-(-2)^5\right)(5)}{1-(-2)} = 55$

13. $a_1 = -2$, $r = \frac{2}{3}$

$a_{10} = -2\left(\frac{2}{3}\right)^9 = \frac{-1024}{19,683}$

15. $\sum\limits_{n=1}^{5} 2(3)^n$, $a_1 = 6$, $r = 3$

$S_5 = \frac{(1-3^5)6}{1-3} = 726$

17. $\displaystyle\sum_{k=1}^{10}\left[\left(\tfrac{1}{2}\right)^k + \left(\tfrac{3}{2}\right)^k\right]$

$$S_{10} = \frac{\left(1-(\tfrac{1}{2})^{10}\right)\left(\tfrac{1}{2}\right)}{1-\tfrac{1}{2}} + \frac{\left(1-(\tfrac{3}{2})^{10}\right)\left(\tfrac{3}{2}\right)}{1-\tfrac{3}{2}}$$

$$= \left(1-\tfrac{1}{2^{10}}\right) + \left(1-\tfrac{3^{10}}{2^{10}}\right)(-3)$$

$$= \frac{3^{11}-1}{2^{10}} - 2$$

19. $\displaystyle\sum_{k=1}^{6} 5\left(\tfrac{1}{2}\right)^k, \quad a_1 = \tfrac{5}{2}, \quad r = \tfrac{1}{2}$

$$S_6 = \left(\frac{1-\left(\tfrac{1}{2}\right)^6}{1-\tfrac{1}{2}}\right)\left(\tfrac{5}{2}\right) = \left(\frac{1-\tfrac{1}{2^6}}{\tfrac{1}{2}}\right)\tfrac{5}{2}$$

$$= \left(1-\tfrac{1}{2^6}\right)5 = \frac{315}{64}$$

21. $\displaystyle\sum_{n=1}^{3}\left(\tfrac{1}{4}\right)^n, \quad a_1 = \tfrac{1}{4}, \quad r = \tfrac{1}{4}$

$$S_3 = \left(\frac{1-\left(\tfrac{1}{4}\right)^3}{1-\tfrac{1}{4}}\right)\left(\tfrac{1}{4}\right) = \left(\frac{1-\tfrac{1}{64}}{\tfrac{3}{4}}\right)\tfrac{1}{4} = \frac{21}{64}$$

23. $r = \tfrac{4}{5}, \quad a_n = \left(\tfrac{4}{5}\right)^n a_1$

$$\tfrac{1}{4}a_1 = \frac{4^n}{5^n}$$

$$5^n = 4^{n+1}$$

$$n \ln 5 = (n+1)\ln 4$$

$$n(\ln 5 - \ln 4) = \ln 4$$

$$n = \frac{\ln 4}{\ln 5 - \ln 4} = 6.212 \Rightarrow 7 \text{ strokes}$$

25. $a_3 = 4, \quad a_4 = 8$

$$r = \tfrac{8}{4} = 2$$

$$a_n = a_1 2^n$$

$$a_3 = 4 = a_1 2^3$$

$$a_1 = \tfrac{1}{2}$$

27. $f(x) = 3^x$

$f(1)=3, \quad f(2)=3^2=9, \quad f(3)=3^3=27, \quad f(4)=3^4=81, \quad f(5)=3^5=243$

29. $f(z) = \left(\frac{1}{2}\right)^x$

$f(1)=\frac{1}{2}, \quad f(2)=\left(\frac{1}{2}\right)^2=\frac{1}{4}, \quad f(3)=\left(\frac{1}{2}\right)^3=\frac{1}{8}, \quad f(4)=\left(\frac{1}{2}\right)^4=\frac{1}{16}, \quad f(5)=\left(\frac{1}{2}\right)^5=\frac{1}{32}$

33. $p = 8000$

$t = 12{,}000$ after 6 years

$t = p(i+r)^6$

$12000 = 8000(i+r)^6$

$1.5 = (i+r)^6$

$i+r = (1.5)^{1/6}$

$r = (1.5)^{1/6} - 1 = .07 = 7\%$

35. Let w_0 be initial weight

$.10w = w_0\left(\frac{1}{2}\right)^n$

$\frac{w_0}{10} = w_0\left(\frac{1}{2}\right)^n$

$\frac{1}{10} = \left(\frac{1}{2}\right)^n$

$\ln .1 = n \ln .5$

$n = 3.32$

Section 9.5   The General Power of a Binomial

1. $(x - y)^6 = x^6 - 6x^5y + 15x^4y^2 - 20x^3y^3 + 15x^2y^4 - 6xy^5 + y^6$

3. $(3x - y)^5 = (3x)^5 - 5(3x)^4y + 10(3x)^3y^2 - 10(3x)^2y^3 + 5(3x)y^4 - y^5$
$= 243x^5 - 405x^4y + 270x^3y^2 - 90x^2y^3 + 15xy^4 - y^5$

5. $\left(x + \frac{1}{x}\right)^5 = x^5 + 5x^4\left(\frac{1}{x}\right) + 10x^3\left(\frac{1}{x}\right)^2 + 10x^2\left(\frac{1}{x}\right)^3 + 5x\left(\frac{1}{x}\right)^4 + \left(\frac{1}{x}\right)^5$
$= x^5 + 5x^3 + 10x + \frac{10}{x} + \frac{5}{x^3} + \frac{1}{x^5}$

7. $(y^2 + 3x)^3 = (y^2)^3 + 3(y^2)^2(3x) + 3y^2(3x)^2 + (3x)^3$
$= y^6 + 9y^4x + 27y^2x^2 + 27x^3$

9. $(\sqrt{x} + \sqrt{y})^6 = (\sqrt{x})^6 + 6(\sqrt{x})^5\sqrt{y} + 15(\sqrt{x})^4(\sqrt{y})^2 + 20(\sqrt{x})^3(\sqrt{y})^3$
$+ 15(\sqrt{x})^2(\sqrt{y})^4 + 6(\sqrt{x})(\sqrt{y})^5 + (\sqrt{y})^6$
$= x^3 + 6x^2\sqrt{xy} + 15x^2y + 20xy\sqrt{xy} + 15xy^2 + 6y^2\sqrt{xy} + y^3$

11. $\left(x^{1/3} + x^{-3/4}\right)^4 = \left(x^{1/3}\right)^4 + 4\left(x^{1/3}\right)^3 x^{-3/4} + 6\left(x^{1/3}\right)^2\left(x^{-3/4}\right)^2$

$$+ 4\left(x^{1/3}\right)\left(x^{-3/4}\right)^3 x^{-3/4} + \left(x^{-3/4}\right)^4$$

$$= x^{4/3} + 4x^{1/4} + 6x^{-5/6} + 4x^{-23/12} + x^{-3}$$

13. $(2+.1)^4 = 2^4 + 4(2)^3(.1) + 6(2)^2(.1)^2 + 4(2)(.1)^3 + (.1)^4$

$$= 16 + 3.2 + .24 + .008 + .0001 = 19.4418$$

15. $\left(x^{-3/4} - x^{3/4}\right)^3 = \left(x^{-3/4}\right)^3 - 3\left(x^{-3/4}\right)^2 x^{3/4} + 3\left(x^{-3/4}\right)\left(x^{3/4}\right)^2 - \left(x^{3/4}\right)^3$

$$= x^{-9/4} - 3x^{-3/4} + 3x^{3/4} - x^{9/4}$$

17. $(\sqrt{x} + \sqrt{y})^{10}$

4th term $= C_{10,\,3}(\sqrt{x})^7(\sqrt{y})^3$

$$= \frac{10!}{3!\,7!}x^3\sqrt{x}\,y\sqrt{y} = 120x^3 y\sqrt{xy}$$

19. $\left(3x^3 - 7y^{1/3}\right)^{20}$

12th term $= C_{20,\,11}(3x^3)^9\left(-7y^{1/3}\right)^{11}$

$$= \frac{20!}{9!\,11!}3^9 x^{27}(-1)7^{11}y^{11/3}$$

$$= \frac{-20!\ 3^9 7^{11}}{9!\ 11!}x^{27}y^{11/3}$$

21. $(x^3 - 2y^2)^5$

$Ax^{12}y^2 = C_{5,\,k}(x^3)^{5-k}(-2y^2)^k$

$12 = 3(5-k), \quad k = 1$

$Ax^{12}y^2 = C_{5,\,1}(x^3)^4(-2y^2)^1$

$Ax^{12}y^2 = \frac{5!}{4!\,1!}x^{12}(-1)2y^2$

$A = \frac{-2 \cdot 5!}{4!} = -10$

23. $\left(x + \frac{1}{3x}\right)^8$

5th term $= C_{8,\,4}x^4\left(\frac{1}{3x}\right)^4$

$$= \frac{8!}{4!\,4!}x^4 \cdot \frac{1}{81x^4} = \frac{70}{81}$$

25. $(2.3)^6 = (2+.3)^6 \simeq 2^6 + 6(2^5)(.3) = 121.66$

27. $(4 - .01)^3 \simeq 4^3 - 3(4)^2(.01) = 63.52$

29. $(1.01)^{500} = (1+.01)^{500} = 1^{500} + C_{500,\,1}1^{499}(.01) = 1 + 500(.01) = 6$

31. $\sqrt{24} = (25-1)^{1/2} = 25^{1/2}\left(1 - \dfrac{1}{25}\right)^{1/2}$

$$= 5(1-.04)^{1/2} \simeq 5\left[1 + \frac{1}{2}(-.04) + \frac{\frac{1}{2}\left(-\frac{1}{2}\right)}{1(2)}(-.04)^2\right] = 4.899$$

33. $\sqrt[3]{7} = (8-1)^{1/3} = 8^{1/3}\left(1-\dfrac{1}{8}\right)^{1/3} = 2(1-.125)^{1/3}$

$$\simeq 2\left[1 + \frac{1}{3}(-.125) + \frac{\left(\frac{1}{3}\right)\left(-\frac{2}{3}\right)(-.125)^2}{1(2)}\right] = 1.913194$$

35. $\sqrt[3]{30} = (27+3)^{1/3} = 27^{1/3}\left(1+\dfrac{1}{9}\right)^{1/3} = 3(1+.1111)^{1/3}$

$$\simeq 3\left[1 + \frac{1}{3}(.1111) + \frac{\left(\frac{1}{3}\right)\left(-\frac{2}{3}\right)(.1111)^2}{(1)(2)}\right] = 3.107$$

37. $\sqrt[4]{80} = (81-1)^{1/4} = 81^{1/4}\left(1-\dfrac{1}{81}\right)^{1/4} = 3(1-.01235)^{1/4}$

$$\simeq 3\left[1 + \frac{1}{4}(-.01235) + \frac{\left(\frac{1}{4}\right)\left(-\frac{3}{4}\right)(-.01235)^2}{(2)(1)}\right] = 2.9907$$

39. $(98)^{1/2} = (100-2)^{1/2} = 100^{1/2}\left(1-\dfrac{2}{100}\right)^{1/2} = 10(1-.02)^{1/2}$

$$\simeq 10\left[1 + \frac{1}{2}(-.02) + \frac{\frac{1}{2}\left(-\frac{1}{2}\right)(-.02)^2}{(1)(2)}\right] = 9.8995$$

1. $P_{7,\,4} = \frac{7!}{3!} = 7 \cdot 6 \cdot 5 \cdot 4 = 840$

   $C_{7,\,4} = \frac{7!}{4!\ 3!} = 35$

3. $P_{n,\,3} = 9P_{n,\,2}$

   $\frac{n!}{(n-3)!} = \frac{9n!}{(n-2)!}$

   $\frac{1}{(n-3)!} = \frac{9}{(n-2)!}$

   $(n-2)! = 9(n-3)!$

   $n-2 = 9$

   $n = 11$

5. $\sum\limits_{k=1}^{4} P_{4,\,k}$

   $= \frac{4!}{3!} + \frac{4!}{2!} + \frac{4!}{1!} + \frac{4!}{0!}$

   $= 4 + 4(3) + 4(3)(2) + 4(3)(2)$

   $= 64 = 4^3$

7. $C_{8,\,3} = \frac{8!}{3!\ 5!} = 56$

9. $P_{4,\,4} = \frac{4!}{0!} = 24$

11. $9 \cdot 8 \cdot 7 \cdot 6 \cdot 5 \cdot 4 \cdot 3 = 181,440$

13. $\frac{13 \cdot 4!}{3!\ 1!} \cdot \frac{12 \cdot 4!}{2!\ 2!} = 13 \cdot 4 \cdot 12 \cdot 3 \cdot 2 = 3744$

15. $5 \cdot 5 \cdot 5 \cdot 2 = 250$

17. $C_{10,\,4} + C_{8,\,2} = \frac{10!}{6!\ 4!} - \frac{8!}{6!\ 2!} = 210 - 28 = 182$

Section 9.7   Probability

1. $P\,(\text{a heart}) = \frac{13}{52} = \frac{1}{4}$

3. $P\,(\text{bolt}) = \frac{3}{8}$

   $P\,(\text{nut}) = \frac{5}{8}$

5. $P\,(\text{odd}) = \frac{3}{6} = \frac{1}{2}$

7. $P\,(\text{2 heads}) = \frac{C_{3,\,2}}{8} = \frac{3!}{2!\ 1!} \cdot \frac{1}{8} = \frac{3}{8}$

9. $P\,(\text{not red}) = \frac{C_{10,\,1}}{C_{13,\,1}} = \frac{10!}{1!\ 9!} \cdot \frac{12!\ 1!}{13!} = \frac{10}{13}$

11. $P \text{ (3 tails)} = \dfrac{C_{6,3}}{2^6} = \dfrac{6!}{3!\ 3!} \cdot \dfrac{1}{64} = \dfrac{5}{16}$

13. $P \text{ (2 nickels)} = \dfrac{C_{15,2}}{C_{26,2}} = \dfrac{15!}{13!\ 2!} \cdot \dfrac{24!\ 2!}{26!} = \dfrac{21}{65}$

15. $P \text{ (vowels)} = \dfrac{C_{14,2}}{C_{11,2}} = \dfrac{4!}{2!\ 2!} \cdot \dfrac{9!\ 2!}{11!} = \dfrac{6}{55}$

17. $P(3) = \dfrac{2}{36} = \dfrac{1}{18}$

19. $P \text{ (exactly two acres)} = \dfrac{C_{4,2}\ C_{48,3}}{C_{52,5}} = \dfrac{4!}{2!\ 2!} \cdot \dfrac{48!}{45!\ 3!} \cdot \dfrac{47!\ 5!}{52!} = \dfrac{2162}{54145}$

Section 9.8   Mathematical Induction

1. $\displaystyle\sum_{k=1}^{n} 2k = n(n+1)$

Let $k = 1$, $\quad 2(1) = 1(1+1)$

$\qquad\qquad 2 = 2 \ \checkmark$

$S_k = k(k+1)$

$S_{k+1} = S_k + 2(k+1)$

$\qquad = k(k+1) + 2(k+1)$

$\qquad = (k+1)(k+2)$

3. $\displaystyle\sum_{k=1}^{n} (3k-2) = \dfrac{n(3-1)}{2}$

Let $k = 1$, $\quad 3-2 = \dfrac{1(3-1)}{2}$

$\qquad\qquad 1 = 1 \ \checkmark$

$S_k = \dfrac{k(3k-1)}{2}$

$S_{k+1} = S_k + \left[3(k+1) - 2 \right]$

$\qquad = \dfrac{k(3k-1)}{2} + 3(k+1) - 2$

$\qquad = \dfrac{3k^2 + 5k + 2}{2}$

$\qquad = \dfrac{(3k+2)(k+1)}{2}$

$\qquad = \dfrac{\left[3(k+1) - 1 \right](k+1)}{2}$

5. $\displaystyle\sum_{k=1}^{n} k(k+1) = \frac{n(n+1)(n+1)}{3}$

Let $k = 1$, $\quad 1(2) = \frac{1(2)(3)}{3}$

$\qquad\qquad\qquad 2 = 2 \checkmark$

$S_k = \dfrac{k(k+1)(k+2)}{3}$

$S_{k+1} = S_k + (k+1)(k+2)$

$\qquad = \dfrac{k(k+1)(k+2)}{3} + (k+1)(k+2)$

$\qquad = (k+1)(k+2)(\tfrac{k}{3}+1)$

$\qquad = \dfrac{(k+1)(k+2)(k+3)}{3}$

7. $\displaystyle\sum_{k=1}^{n} k^2 = \frac{n(n+1)(2n+1)}{6}$

Let $k = 1$, $\qquad 1 = \dfrac{1(2)(3)}{6}$

$\qquad\qquad\qquad 1 = 1 \checkmark$

$S_k = \dfrac{k(k+1)(2k+1)}{6}$

$S_{k+1} = S_k + (k+1)^2$

$\qquad = \dfrac{k(k+1)(2k+1) + (k+1)^2}{6}$

$\qquad = \dfrac{(k+1)\Big(k(2k+1) + 6(k+1)\Big)}{6}$

$\qquad = \dfrac{(k+1)(2k^2 - 7k + 6)}{6}$

$\qquad = \dfrac{(k+1)(2k+3)(k+2)}{6}$

$\qquad = \dfrac{(k+1)(k+2)\Big(2(k+1) + 1\Big)}{6}$

9. $\displaystyle\sum_{k=1}^{n} r^{k-1} = \frac{1-r^n}{1-r}$, $\qquad r \neq 1$

Let $k = 1$, $\qquad r^\circ = \dfrac{1-r}{1-r}$

$\qquad\qquad\qquad 1 = 1 \checkmark$

$S_k = \dfrac{1-r^k}{1-r}$

$S_{k+1} = S_k + r^{(k+1)-1}$

$\qquad = \dfrac{1-r^k}{1-r} + r^k$

$\qquad = \dfrac{1 - r^k + r^k(1-r)}{1-r}$

$\qquad = \dfrac{1 - r^k + r^k - r^{k+1}}{1-r}$

$\qquad = \dfrac{1 - r^{k+1}}{1-r}$

11. $n(n+1)(n+2) = 6c,$ c is an integer

Let $n = 1, 1(2)(3) = 6(1)$

$$6 = 6 \checkmark$$

Assume true for k.

$k(k+1)(k+2) = 6c$, for some integer c

Consider $(k+1)(k+2)(k+3)$

$$= k^3 + 6k^2 + 11k + 6$$
$$= k^3 + 3k^2 + 3k^2 + 9k + 2k + 6$$
$$= (k^3 + 3k^2 + 2k) + (3k^2 + 9k + 6)$$
$$= k(k^2 + 3k + 2) + 3(k^2 + 3k + 2)$$
$$= k(k+1)(k+2) + 3(k+1)(k+2)$$

Let $k =$ odd number

Then $k+1 =$ even

$$k+1 = 2d \text{ for some integer } d$$

So $(k+1)(k+2)(k+3) = 6c + 3(2d)(k+2)$
$$= 6(c+d)(k+2),$$

c, d, $k+1$ are integers

so $(c+d)(k+1)$ is an integer

Let $k =$ even number

Then $k+2 =$ even

$$k+2 = 2d \text{ for some integer } d$$

So $(k+1)(k+2)(k+3) = 6c + 3(2d)(k+1)$
$$= 6(c+d)(k+1)$$

c, d, $k+1$ are integers

so $(c+d)(k+1)$ is an integer

13. $\displaystyle\sum_{k=1}^{n+1}\frac{1}{n+k} \leq \frac{5}{6}$

Let $n = 1$, $\quad\displaystyle\sum_{k=1}^{2}\frac{1}{1+k} = \frac{1}{1+1} + \frac{1}{1+2} \leq \frac{5}{6}$

Assume true for $n = r$

$$n = r \Rightarrow \sum_{k=1}^{r+1}\frac{1}{r+k} = \frac{1}{r+1} + \frac{1}{r+2} + \cdots + \frac{1}{r+r+1} \leq \frac{5}{6}$$

Consider $\displaystyle\sum_{k=1}^{(r+1)+1}\frac{1}{(r+1)+k} = \sum_{k=1}^{r+2}\frac{1}{r+1+k}$

$$= \frac{1}{r+2} + \frac{1}{r+3} + \cdots + \frac{1}{(r+1)+(r+2)}$$

$$= \frac{1}{r+2} + \frac{1}{r+3} + \cdots + \frac{1}{2r+3}$$

$$= \frac{1}{r+2} + \frac{1}{r+3} + \cdots + \frac{1}{2r+1} + \left(\frac{1}{2r+2} + \frac{1}{2r+3}\right)$$

$$\leq \frac{1}{r+2} + \frac{1}{r+3} + \cdots + \frac{1}{2r+1} + \frac{1}{r+1}$$

$$= \sum_{k=1}^{r+1}\frac{1}{r+k} \leq \frac{5}{6}$$

Chapter 9   Review

11. $a_1 = 1,\quad d = 2$

$S_{20} = 1(20) + \dfrac{20(19)(2)}{2} = 400$

13. $a_1 = 3,\quad d = 6$

$S_{20} = 3(20) + \dfrac{20(19)(6)}{2} = 1200$

15. $a_1 = 1,\quad d = 0$

$S_{20} = 1(20) + 0 = 20$

17. $a_1 = 2,\quad r = 2$

$S_{20} = \dfrac{(1-2^{20})2}{1-2} = (1-2^{20})(-2) = -2 + 2^{21} = 2{,}097{,}150$

19. $a_1 = -5,\quad r = -5$

$S_{20} = \dfrac{\left(1 - (-5)^{20}\right)(-5)}{1 - (-5)} = \dfrac{(1-5^{20})(-5)}{6} = \dfrac{-5 + 5^{21}}{6} = 7.947\times10^{13}$

21. $\displaystyle\sum_{k=1}^{2} 2^k = 2 + 2^2 = 6$

23. $\displaystyle\sum_{i=1}^{4} i(i+2) = 1(3) + 2(4) + 3(5) + 4(6) = 50$

25. $\displaystyle\sum_{k=2}^{10}\left[\frac{1}{k+3} - \frac{1}{k+2}\right] = \left(\frac{1}{5} - \frac{1}{4}\right) + \left(\frac{1}{6} - \frac{1}{5}\right) + \left(\frac{1}{7} - \frac{1}{6}\right) + \left(\frac{1}{8} - \frac{1}{7}\right) + \left(\frac{1}{9} - \frac{1}{8}\right)$
$$+ \left(\frac{1}{10} - \frac{1}{9}\right) + \left(\frac{1}{11} - \frac{1}{10}\right) + \left(\frac{1}{12} - \frac{1}{11}\right) + \left(\frac{1}{13} - \frac{1}{12}\right)$$
$$= -\frac{1}{4} + \frac{1}{13} = -\frac{9}{52}$$

27. $\displaystyle\sum_{k=2}^{10} a_k = a_2 + a_3 + a_4 + a_5 + a_6 + a_7 + a_8 + a_9 + a_{10}$

$\displaystyle\sum_{k=4}^{12} a_{k-2} = a_2 + a_3 + a_4 + a_5 + a_6 + a_7 + a_8 + a_9 + a_{10}, \quad$ True

29. $\displaystyle\sum_{t=1}^{5} (t^2-1)(t-2) = 0(-1) + (3)(0) + (8)(1) + (15)(2) + (24)(3) = 110$

$\displaystyle\sum_{s=-1}^{3} (s^2-1)(s) = (2)(-1) + (1)(0) + (2)(1) + (5)(2) + (10)(3) = 40, \quad$ False

31. $\left(x - \frac{1}{x}\right)^{10}$

$C_{10,\,k}\, x^k\left(-\frac{1}{x}\right)^{10-k} = C_{10,\,k}\,\frac{x^k}{x^{10-k}}(-1)^{10-k}$

$\dfrac{x^k}{x^{10-k}} = 1$

$x^{k-(10-k)} = x^{\circ}$

$k - (10-k) = 0$

$\qquad k = 5$

$C_{10,\,5}\, x^5\left(-\frac{1}{x}\right)^5 = \frac{10!}{5!\,5!}(-1)^5 = -252$

33. $(x-5y)^4 = x^4 - 4x^3(5y) + 6x^2(5y)^2 - 4x(5y)^3 + (5y)^4$
$$= x^4 - 20x^3y + 150x^2y^2 - 500xy^3 + 625y^4$$

35. $(1.01)^{100} = (1+.01)^{100} \simeq 1^{100} + C_{100,\,1}1^{99}(.01) + C_{100,\,2}1^{98}(.01)^2$

$$= 1 + 100(.01) + \frac{100!}{98!\,2!}1(.01)^2 = 2.495$$

37. $2^{20} = 1,048,576$

39. $P_{5,\,2} = \frac{5!}{3!} = 20$

$$P_{100,\,2} = \frac{100!}{98!} = 9900$$

41. $\dfrac{(2n)!}{(2n-3)!} = \dfrac{(2n)(2n-1)(2n-2)\cancel{(2n-3)}\cancel{(2n-4)}\cdots\cancel{1}}{\cancel{(2n-3)}\cancel{(2n-4)}\cdots\cancel{1}}$

$$= 2n(2n-1)(2n-2)$$

47. $P(4 \text{ tails}) = \dfrac{C_{7,\,4}}{2^7} = \dfrac{7!}{4!\,3!} \cdot \dfrac{1}{2^7} = \dfrac{35}{128}$

Chapter 10 The Conic Sections

Section 10.1   Conic Sections: The Parabola

1. $y^2 = -8x$

    $4a = -8, \quad a = -2$

    Focus: $(-2, 0)$

    Directrix: $x = 2$

    Right Chord: $(-2, -4), (-2, 4)$

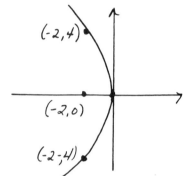

3. $2x^2 = 12y$

    $x^2 = 6y$

    $4a = 6, \quad a = \frac{3}{2}$

    Focus: $(0, \frac{3}{2})$

    Directrix: $y = -\frac{3}{2}$

    Right Chord: $(-3, \frac{3}{2}), (3, \frac{3}{2})$

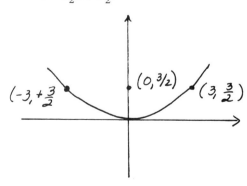

5. $y^2 + 16x = 0$

$y^2 = -16x$

$4a = -16, \quad a = -4$

Focus: $(-4, 0)$

Directrix: $x = 4$

Right Chord: $(-4, -8), (-4, 8)$

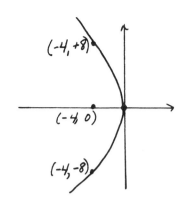

7. $y^2 = 3x$

$4a = 3, \quad a = \frac{3}{4}$

Focus: $\left(\frac{3}{4}, 0\right)$

Directrix: $x = -\frac{3}{4}$

Right Chord: $\left(\frac{3}{4}, \frac{3}{2}\right), \left(\frac{3}{4}, -\frac{3}{2}\right)$

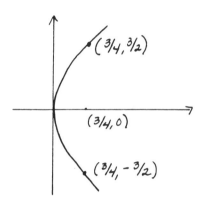

9. $y^2 = -2x$

$4a = -2, \quad a = -\frac{1}{2}$

Focus: $\left(-\frac{1}{2}, 0\right)$

Directrix: $x = \frac{1}{2}$

Right Chord: $\left(-\frac{1}{2}, -1\right), \left(-\frac{1}{2}, 1\right)$

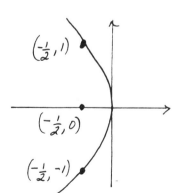

11. Focus: $(0, 2)$

   Directrix: $y = -2$

   $a = 2$

   $x^2 = 4ay, \quad x^2 = 8y$

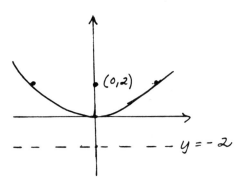

13. Focus: $\left(\frac{3}{2}, 0\right)$

   Directrix: $x = -\frac{3}{2}$

   $a = \frac{3}{2}$

   $y^2 = 4ax, \quad y^2 = 6x$

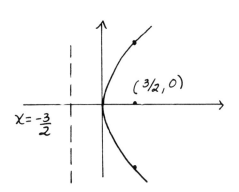

15. Right Chord: $(2, -1), (-2, -1)$

   Vertex: $(0, 0)$

   Focus: $(0, -1)$

   $a = -1$

   $x^2 = 4ay, \quad x^2 = -4y$

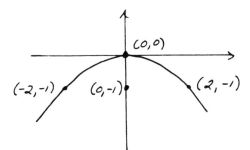

17. Vertex: (0, 0)

Vertical Axis → $x^2 = 4ay$

(2, 4)

$4 = 4a(4)$

$a = \frac{1}{4}$

$x^2 = 4\left(\frac{1}{4}\right)y = y$

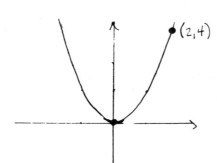

19.

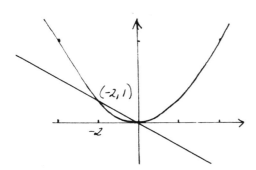

21.

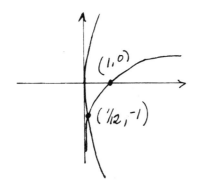

23. $x^2 = 4ay$

$250{,}000 = 4a(250)$

$4a = 1000$

$x^2 = 1000y$

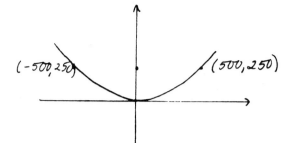

Section 10.2   The Ellipse and the Circle

1. $5x^2 + y^2 = 25$

    $\dfrac{x^2}{5} + \dfrac{y^2}{25} = 1$

    $a = 5, \quad b = \sqrt{5}$

    $c = \sqrt{25-5} = \pm 2\sqrt{5}$

    Focus: $(0, \pm 2\sqrt{5})$

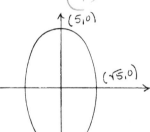

3. $16x^2 + 4y^2 = 16$

    $x^2 + \dfrac{y^2}{4} = 1$

    $a = 2, \quad b = 1$

    $c = \sqrt{4-1} = \pm \sqrt{3}$

    Focus: $(0, \pm \sqrt{3})$

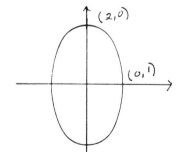

5. $x^2 + y^2 = 9$

$r = 3$

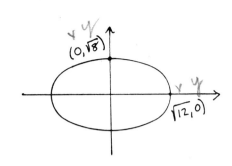

7. $2x^2 + 3y^2 = 24$

$\frac{x^2}{12} + \frac{y^2}{8} = 1$

$a = 2\sqrt{3}, \quad b = 2\sqrt{2}$

$c = \sqrt{12-8} = 2$

Focus: $(\pm 2, 0)$

9. $9x^2 + 4y^2 = 4$

$\frac{x^2}{\frac{4}{9}} + y^2 = 1$

$a = 1, \quad b = \frac{2}{3}$

$c = \sqrt{1-\frac{4}{9}} = \pm\frac{\sqrt{5}}{3}$

Focus: $(0, \pm\frac{\sqrt{5}}{3})$

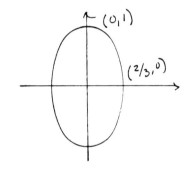

11. Vertices: $(\pm 4, 0)$

Minor Axis: 6

$\frac{x^2}{16} + \frac{y^2}{9} = 1$

$9x^2 + 16y^2 = 144$

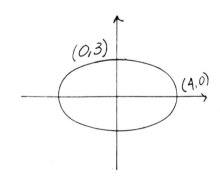

13. Vertices: $(0, \pm 5)$

Semi-minor Axis: $\frac{3}{2}$

$$\frac{x^2}{\frac{9}{4}} + \frac{y^2}{25} = 1$$

$$9y^2 + 100x^2 = 225$$

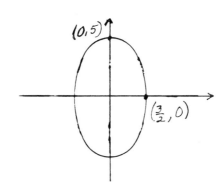

15. Major Axis: 10

Foci: $(\pm 4, 0)$

$a = 5, \quad c = 4$

$b^2 = 25 - 16 = 9, \quad b = 3$

$$\frac{x^2}{25} + \frac{y^2}{9} = 1$$

$$9x^2 + 25y^2 = 225$$

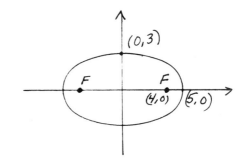

17. Foci: $(\pm 1, 0)$

Major Axis: 8

$c = 1, \quad a = 4$

$b^2 = 4^2 - 1 = 15, \quad b = \sqrt{15}$

$$\frac{x^2}{16} + \frac{y^2}{15} = 1$$

$$15x^2 + 16y^2 = 240$$

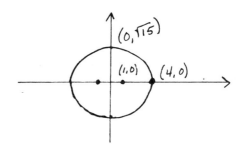

19. Vertices: $\left(\pm\frac{5}{2}, 0\right)$

$$\frac{x^2}{\frac{25}{4}} + \frac{y^2}{b^2} = 1$$

$$\frac{4x^2}{25} + \frac{y^2}{b^2} = 1$$

If $(1, 1)$, then

$$\frac{4}{25} + \frac{1}{b^2} = 1$$

$$b^2 = \frac{25}{21}, \quad b = \frac{5}{\sqrt{21}}$$

$$\frac{4x^2}{25} + \frac{21y^2}{25} = 1$$

$$4x^2 + 21y^2 = 25$$

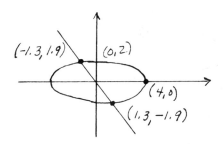

21.

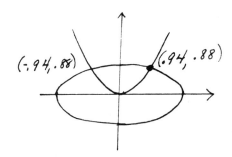

23.

– 181 –

25. Major Axis: 10

   Minor Axis: 3

$$\frac{x^2}{25} + \frac{y^2}{\left(\frac{3}{2}\right)^2} = 1$$

$$9x^2 + 100y^2 = 225$$

Section 10.3   The Hyperbola

1. $x^2 - y^2 = 16$

   $$\frac{x^2}{16} - \frac{y^2}{16} = 1$$

   $a = 4, \quad b = 4$

   $c = \sqrt{16+16} = \sqrt{32}$

   Foci: $\left(\pm\sqrt{32},\, 0\right)$

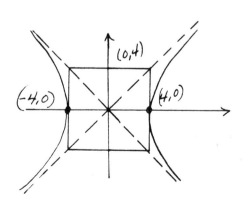

3. $4x^2 - 9y^2 = 36$

$\dfrac{x^2}{9} - \dfrac{y^2}{4} = 1$

$a = 3, \quad b = 2$

$c = \sqrt{9+4} = \sqrt{13}$

Foci: $\left(\pm\sqrt{13}, 0\right)$

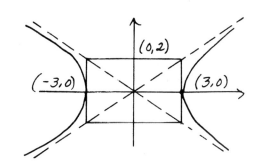

5. $4y^2 - 25x^2 = 100$

$\dfrac{y^2}{25} - \dfrac{x^2}{4} = 1$

$a = 5, \quad b = 2$

$c = \sqrt{25+4} = \sqrt{29}$

Foci: $\left(0, \pm\sqrt{29}\right)$

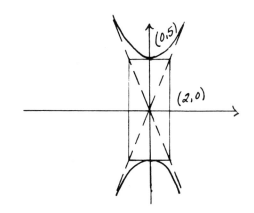

7. $4x^2 - 16y^2 = 25$

$\dfrac{x^2}{\frac{25}{4}} - \dfrac{y^2}{\frac{25}{16}} = 1$

$a = \dfrac{5}{2}, \quad b = \dfrac{5}{4}$

$c = \sqrt{\dfrac{25}{4} + \dfrac{25}{16}} = \dfrac{5}{4}\sqrt{5}$

Foci: $\left(\pm\dfrac{5}{4}\sqrt{5}, 0\right)$

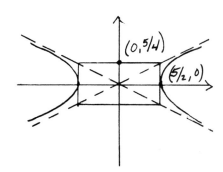

9. $y^2 + 1 = x^2$

$x^2 - y^2 = 1$

$a = 1, \quad b = 1$

$c = \sqrt{1+1} = \sqrt{2}$

Foci: $\left(\pm\sqrt{2}, 0 \right)$

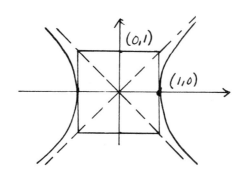

11. Vertices: $(\pm 4, 0)$

Foci: $(\pm 5, 0)$

$\dfrac{x^2}{a^2} - \dfrac{y^2}{b^2} = 1$

$a = 4, \quad c = 5$

$b = \sqrt{25-16} = 3$

$\dfrac{x^2}{16} - \dfrac{y^2}{9} = 1$

$9x^2 - 16y^2 = 144$

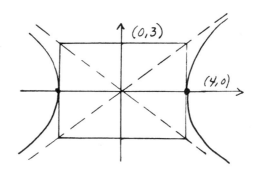

13. Conjugate Axis: 4

Vertices: $(0, \pm 1)$

$\dfrac{y^2}{a^2} - \dfrac{x^2}{b^2} = 1$

$a = 1, \quad b = 2$

$y^2 - \dfrac{x^2}{4} = 1$

$4y^2 - x^2 = 4$

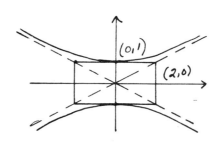

15. Transverse Axis: 6

   Foci: $\left(\pm\frac{7}{2}, 0\right)$

   $\frac{x^2}{a^2} - \frac{y^2}{b^2} = 1$

   $a = 3, \quad c = \frac{7}{2}$

   $b = \sqrt{\frac{49}{4} - 9} = \frac{\sqrt{13}}{2}$

   $\frac{x^2}{9} - \frac{4y^2}{13} = 1$

   $13x^2 - 36y^2 = 117$

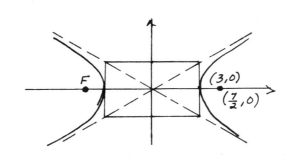

17. Vertices: $(0, \pm 4)$

   Asymptotes: $y = \pm\frac{x}{2}$

   $\frac{y^2}{a^2} - \frac{x^2}{b^2} = 1$

   $a = 4$

   $\frac{1}{2} = \frac{a}{b}, \quad b = 8$

   $\frac{y^2}{16} - \frac{x^2}{64} = 1$

   $4y^2 - 4x^2 = 64$

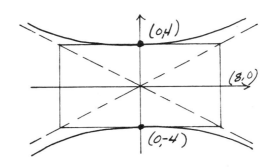

19. Vertices: $(0, \pm 3)$

   $\frac{y^2}{a^2} - \frac{x^2}{b^2} = 1$

   $a = 3$

   $\frac{y^2}{9} - \frac{x^2}{b^2} = 1$

   If $(2, 7)$, then

   $\frac{49}{9} - \frac{4}{b^2} = 1, \quad b^2 = \frac{9}{10}$

   $\frac{y^2}{9} - \frac{10x^2}{9} = 1$

   $y^2 - 10x^2 = 9$

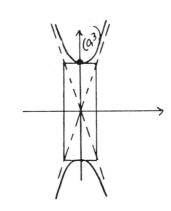

21. $\dfrac{x^2}{\sin^2 C} - \dfrac{y^2}{\cos^2 C} = 1$

$a^2 = \sin^2 C$

$b^2 = \cos^2 C$

$c^2 = \sin^2 C + \cos^2 C$

$c^2 = 1$

$c = \pm 1$

Foci: $(\pm 1, 0)$

Section **10.4**   Translation of Axes

1. **Vertex:** (3, 1)

   Focus: (5, 1)

   $a = 5 - 3 = 2$

   $(y-k)^2 = 4a(x-h)$

   $(y-1)^2 = 8(x-3)$

   $8x = y^2 - 2y + 25$

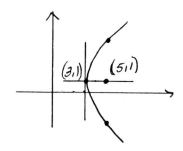

3. **Directrix:** $y = 2$

   Vertex: (1, -1)

   $a = -1 - 2 = -3$

   $(x-h)^2 = 4a(y-k)$

   $(x-1)^2 = -12(y+1)$

   $x^2 - 2x + 13 = -12y$

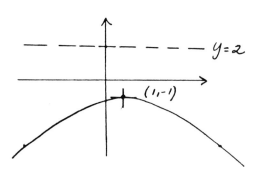

5. Right Chord: $(2, 4)$, $(2, 0)$

$4a = 4$, $a = 1$

Axis: $y = 2$

Vertex: $(1, 2)$

$(y-k)^2 = 4a(x-h)$

$(y-2)^2 = 4(x-1)$

$y^2 - 4y + 8 = 4x$

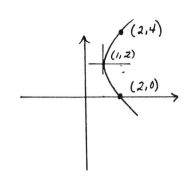

7. Major Axis: 8

Foci: $(5, 1)$, $(-1, 1)$

Center: $(2, 1)$

$a = 4$, $c = 3$,

$b = \sqrt{16-9} = \sqrt{17}$

$$\frac{(x-h)^2}{a^2} + \frac{(y-k)^2}{b^2} = 1$$

$$\frac{(x-2)^2}{16} + \frac{(y-1)^2}{9} = 1$$

$7x^2 - 28x + 16y^2 - 32y - 68 = 0$

$a^2 + b^2 = c^2$

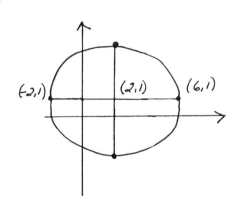

9. Minor Axis: 2

Vertices: $\left(\frac{1}{2}, 0\right)$, $\left(\frac{1}{2}, -8\right)$

Center: $\left(\frac{1}{2}, -4\right)$

$a = 4$, $b = 1$, $c = \sqrt{16+1} = \sqrt{17}$

$$\frac{(x-h)^2}{b^2} + \frac{(y-k)^2}{a^2} = \quad = 1$$

$$\left(x - \frac{1}{2}\right)^2 + \frac{(y+4)^2}{16} = 1$$

$16x^2 - 16x + y^2 + 8y + 4 = 0$

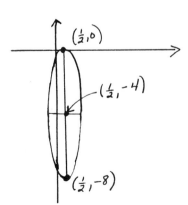

11. Vertices: $(-6, 3)$, $(-2, 3)$

Foci: $(-5, 3)$, $(-3, 3)$

Center: $(-4, 3)$

$a = 2, \quad c = 1, \quad b^2 = 4-1 = 3$

$$\frac{(x-h)^2}{a^2} + \frac{(y-k)^2}{b^2} = 1$$

$$\frac{(x+4)^2}{4} + \frac{(y-3)^2}{3} = 1$$

$3x^2 + 24x + 4y^2 - 24y + 72 = 0$

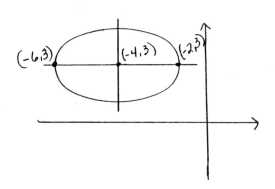

13. Center: $(-1, 2)$

Transverse Axis: 7

Conjugate Axis: 8

Vertical Axis

$a = \frac{7}{2} \qquad b = 4$

$$\frac{(y-k)^2}{a^2} - \frac{(x-h)^2}{b^2} = 1$$

$$\frac{(y-2)^2}{\left(\frac{7}{2}\right)^2} - \frac{(x+1)^2}{4^2} = 1$$

$64y^2 - 49x^2 - 98x - 256y - 577 = 0$

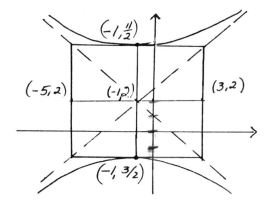

15. Vertices: $(5, 1)$, $(-1, 1)$

Foci: $(6, 1)$, $(-2, 1)$

Center: $(2, 1)$

$a = 3, \quad c = 4$

$b^2 = 16-9 = 7$

$$\frac{(x-h)^2}{a^2} - \frac{(y-k)^2}{b^2} = 1$$

$$\frac{(x-2)^2}{9} - \frac{(y-1)^2}{7} = 1$$

$7x^2 - 28x - 9y^2 + 18y - 44 = 0$

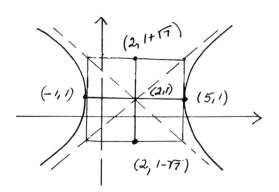

17. Vertices: $(-4, -2), (0, -2)$

Slope of Asymptotes: $\pm\frac{1}{2}$

Center: $(-2, -2)$

$a = 2, \qquad m = \pm\frac{1}{2}$

$\qquad\qquad m = \pm\frac{b}{a}, \quad b = 1$

$\dfrac{(x-h)^2}{a^2} - \dfrac{(y-k)^2}{b^2} = 1$

$\dfrac{(x+2)^2}{4} - \dfrac{(y+2)^2}{1} = 1$

$x^2 + 4x - 4y^2 - 16y - 16 = 0$

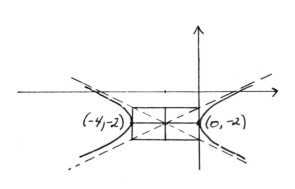

$(-4, -2) \qquad (0, -2)$

19. Center: $(h, 0)$

$(x-h)^2 + (y-0)^2 = r^2$

$(x-h)^2 + y^2 = r^2$

21. Vertex: $(h, 0)$

$(y-k)^2 = 4a(x-h)$

$y^2 = 4a(x-h)$

23. Center: $(h, 0)$

$(x-h)^2 + y^2 = r^2$

If $(0, 0)$, then

$(-h)^2 + 0 = r^2$

$\qquad\qquad h = r$

$(x-r)^2 + y^2 = r^2$

25.

$y = 20x - \dfrac{1}{10}x^2$

$10y = 200x - x^2$

$x^2 - 200x = -10y$

$x^2 - 200x + (100)^2 = -10y + (100)^2$

$(x-100)^2 = -10(y-1000)$

Section 10.5   Rotation of Axes

1. $x' = 3\cos 30° + 30° = \dfrac{3\sqrt{3} + 1}{2}$

$y' = -3\sin 30° + \cos 30° = \dfrac{-3 + \sqrt{3}}{2}$

3. $x' = -\cos 30° - \sin 30° = \dfrac{-\sqrt{3} - 1}{2}$

$y' = \sin 30° - \cos 30° = \dfrac{1 - \sqrt{3}}{2}$

5. $x' = 2 \cos 30° - \sin 30° = \dfrac{2\sqrt{3} - 1}{2}$

   $y' = -2 \sin 30° - \cos 30° = \dfrac{-2 - \sqrt{3}}{2}$

7. $x = \dfrac{x'}{\sqrt{2}} - \dfrac{y'}{\sqrt{2}}$

   $y = \dfrac{x'}{\sqrt{2}} + \dfrac{y'}{\sqrt{2}}$

   $y = 2x$

   $\dfrac{x' + y'}{\sqrt{2}} = \dfrac{2(x' - y')}{\sqrt{2}}$

   $3y' = x'$

   $y' = \tfrac{1}{3}x'$

9. $x = x' \cos 60° - y' \sin 60° = \dfrac{x'}{2} - \dfrac{\sqrt{3}y'}{2}$

   $y = x' \sin 60° + y' \cos 60° = \dfrac{\sqrt{3}x'}{2} + \dfrac{y'}{2}$

   $2x + 7y = 3$

   $\dfrac{2\left(x' - \sqrt{3}y'\right)}{2} + \dfrac{7\left(\sqrt{3}x' + y'\right)}{2} = 3$

   $\left(2 + 7\sqrt{3}\right)x' + \left(7 - 2\sqrt{3}\right)y' = 6$

11. $x = x' \cos 90° - y' \sin 90° = -y'$

    $y = x' \sin 90° + y' \cos 90° = x'$

    $x^2 + 4y^2 = 4$

    $(-y')^2 + 4(x')^2 = 4$

    $y'^2 + 4x'^2 = 4$

13. $x = x' \cos 30° - y' \sin 30° = \dfrac{\sqrt{3}x'}{2} - \dfrac{y'}{2}$

$y = x' \sin 30° + y' \cos 30° = \dfrac{x'}{2} + \dfrac{\sqrt{3}y'}{2}$

$$x^2 + 2x + y^2 = 0$$

$$\dfrac{\left(\sqrt{3}x' - y'\right)^2}{4} + \dfrac{2\left(\sqrt{3}x' - y'\right)}{2} + \dfrac{\left(x' + \sqrt{3}y'\right)^2}{4} = 0$$

$$\dfrac{3x'^2}{4} - \dfrac{\sqrt{3}x'y'}{2} + \dfrac{y'^2}{4} + \sqrt{3}x' - y' + \dfrac{x'^2}{4} + \dfrac{\sqrt{3}y'x'}{2} + \dfrac{3y'^2}{4} = 0$$

$$x'^2 + y'^2 + \sqrt{3}x' - y' = 0$$

15. $\cos \theta = .6$

$\theta = 53.13°$

17. $x = \dfrac{x'}{\sqrt{2}} - \dfrac{y'}{\sqrt{2}}$

$y = \dfrac{x'}{\sqrt{2}} + \dfrac{y'}{\sqrt{2}}$

$$3x^2 - 2xy + 3y^2 = 2$$

$$\dfrac{3(x'-y')^2}{2} - \dfrac{2(x'-y')(x'+y')}{\sqrt{2}\sqrt{2}} + \dfrac{3(x'+y')^2}{2} = 2$$

$$3(x'^2 - 2x'y' + y'^2) - 2(x'^2 - y'^2) + 3(x'^2 + 2x'y' + y'^2) = 4$$

$$4x'^2 + 8y'^2 = 4$$

$$x'^2 + 2y'^2 = 1$$

19. $x^2 + xy = 1$

$$(x' \cos \theta - y' \sin \theta)^2 + (x' \cos \theta - y' \sin \theta)(x' \sin \theta + y' \cos \theta) = 1$$

$$x'^2(\cos^2 \theta + \sin \theta \cos \theta) + y'^2(\sin^2 \theta - \cos \theta \sin \theta) +$$
$$x'y'(-2 \cos \theta \sin \theta - \sin^2 \theta + \cos^2 \theta) = 1$$

Let $-2 \cos \theta \sin \theta - \sin^2 \theta + \cos^2 \theta = 0$

$$\cos^2 \theta - (1 - \cos^2 \theta) = 2 \cos \theta \sqrt{1 - \cos^2 \theta}$$

$$(2 \cos^2 \theta - 1)^2 = \left(2 \cos \theta \sqrt{1 - \cos^2 \theta}\right)^2$$

$$4 \cos^4 \theta - 4 \cos^2 \theta + 1 = 4 \cos^2 \theta (1 - \cos^2 \theta)$$

$$8 \cos^4 \theta - 8 \cos^2 \theta + 1 = 0$$

$$\cos^2 \theta = \frac{8 \pm \sqrt{64 - 32}}{16} = \frac{2 \pm \sqrt{2}}{4}$$

$$\cos \theta = .9238, \quad \theta = 22.5°$$

$$x'^2(.8534 + .3535) + x'y'(-.707 - .1465 + .8534) + y'^2(.1465 - .3535) =$$

$$\frac{x'^2}{.828} - \frac{y'^2}{4.828} = 1$$

21. $x^2 - xy + y^2 = 1$

$$(x' \cos \theta - y' \sin \theta)^2 - (x' \cos \theta - y' \sin \theta)(x' \sin \theta + y' \cos \theta) +$$
$$(x' \sin \theta + y' \cos \theta)^2 = 1$$

$$x'^2(\cos^2 \theta - \cos \theta \sin \theta + \sin^2 \theta) + y'^2(\sin^2 \theta + \sin \theta \cos \theta + \sin^2 \theta) +$$
$$x'y'(\sin^2 \theta - \cos^2 \theta) = 1$$

Let $\sin^2 \theta - \cos^2 \theta = 0$

$$\sin^2 \theta = \cos^2 \theta$$

$$\theta = 45°$$

$$x'^2\left(1 - \tfrac{1}{2}\right) + y'^2\left(1 + \tfrac{1}{2}\right) = 1$$

$$\frac{x'^2}{2} + \frac{y'^2}{\frac{2}{3}} = 1$$

1. $x^2 + y^2 + 4x + 6y + 4 = 0$

$\qquad x^2 + 4x + y^2 + 6y + 9 = -4 + 4 + 9$

$\qquad\quad (x+2)^2 + (y+3)^2 = 9$

center: $(-2, -3)$,   radius: 3

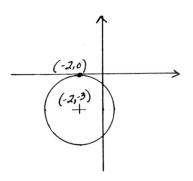

3. $9x^2 + 4y^2 + 18x + 8y - 23 = 0$

$\qquad 9(x^2+2x+1) + 4(y^2+2y+1) = 23 + 9 + 4$

$\qquad\quad 9(x+1)^2 + 4(y+1)^2 = 36$

$\qquad\quad \dfrac{(x+1)^2}{4} + \dfrac{(y+1)^2}{9} = 1$

center: $(-1, -1)$

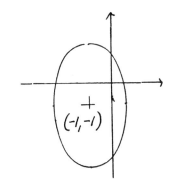

5. $x^2 - y^2 - 4x - 21 = 0$

$\qquad x^2 - 4x + 4 - y^2 = 21 + 4$

$\qquad\quad (x-2)^2 - y^2 = 25$

center: $(2, 0)$,   radius: 5

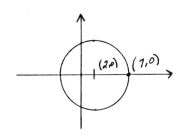

7. $x^2 - 6x - 3y = 0$

$\quad x^2 - 6x + 9 = 3y + 9$

$\quad\quad (x-3)^2 = 3(y+3)$

$\quad 4a = 3, \quad a = \frac{3}{4}$

Vertex: $(3, -3)$

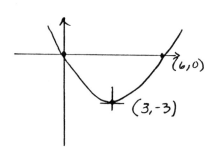

9. $2y^2 - 2y + x - 1 = 0$

$\quad 2\left(y^2 - y + \frac{1}{4}\right) = -x + 1 + \frac{1}{2}$

$\quad\quad 2\left(y - \frac{1}{2}\right)^2 = -\left(x - \frac{3}{2}\right)$

$\quad\quad \left(y - \frac{1}{2}\right)^2 = -\frac{1}{2}\left(x - \frac{3}{2}\right)$

$\quad 4a = -\frac{1}{2}, \quad a = -\frac{1}{8}$

Vertex: $\left(\frac{3}{2}, \frac{1}{2}\right)$

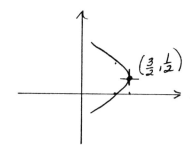

11. $\quad\quad 4x^2 - y^2 + 8x + 2y - 1 = 0$

$\quad 4(x^2 + 2x + 1) - (y^2 - 2y + 1) = 1 + 4 - 1$

$\quad\quad 4(x + 1)^2 - (y-1)^2 = 4$

$\quad\quad (x + 1)^2 - \frac{(y - 1)^2}{4} = 1$

center: $(-1, 1)$

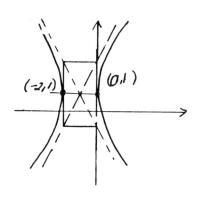

13. $\quad 9x^2 + 4y^2 - 18x + 16y - 11 = 0$

$\quad 9(x^2 - 2x + 1) + 4(y^2 + 4y + 4) = 11 + 9 + 16$

$\quad\quad 9(x - 1)^2 + 4(y + 2)^2 = 36$

$$\frac{(x - 1)^2}{4} + \frac{(y + 2)^2}{9} = 1$$

center: $(1, -2)$

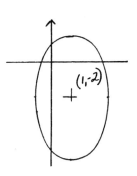

15. $\quad x^2 - 2y^2 + 2y = 0$

$\quad x^2 - 2\left(y^2 - y + \frac{1}{4}\right) = -\frac{1}{2}$

$\quad\quad x^2 - 2\left(y - \frac{1}{2}\right)^2 = -\frac{1}{2}$

$\quad\quad -2x^2 + 4\left(y - \frac{1}{2}\right)^2 = 1$

$$-\frac{x^2}{\frac{1}{2}} + \frac{\left(y - \frac{1}{2}\right)^2}{\frac{1}{4}} = 1$$

center: $\left(0, \frac{1}{2}\right)$

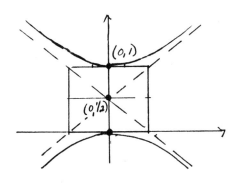

17. $\quad x(x + 4) = y^2 + 3$

$\quad x^2 + 4x - y^2 = 3$

$\quad x^2 + 4x + 4 - y^2 = 7$

$$\frac{(x + 2)^2}{7} - \frac{y^2}{7} = 1$$

center: $(-2, 0)$

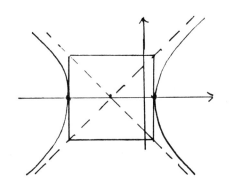

19.
$$y = x^2 + 5x + 7$$
$$y - 7 = x^2 + 5x$$
$$y - 7 + \frac{25}{4} = x^2 + 5x + \frac{25}{4}$$
$$y - \frac{3}{4} = \left(x + \frac{5}{2}\right)^2$$
$$4a = 1, \quad a = \frac{1}{4}$$
vertex: $\left(-\frac{5}{2}, \frac{3}{4}\right)$

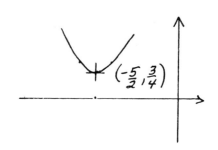

21. $5x^2 + 12xy = 9$

$A = 5, \quad B = 12, \quad C = 0$

$\cot 2\theta = \frac{5-0}{12} = \frac{5}{12}$

$\cos 2\theta = \frac{5}{13}$

$\sin \theta = \sqrt{\frac{1 - \frac{5}{13}}{2}} = \frac{2}{\sqrt{13}}$

$\cos \theta = \sqrt{\frac{1 + \frac{5}{13}}{2}} = \frac{3}{\sqrt{13}}, \ \theta = 33.69°$

$x = x' \frac{3}{\sqrt{13}} - y' \frac{2}{\sqrt{13}}$

$y = x' \frac{2}{\sqrt{13}} + y' \frac{3}{\sqrt{13}}$

$$5\left(\frac{3x' - 2y'}{\sqrt{13}}\right)^2 + 12\left(\frac{3x' - 2y'}{\sqrt{13}}\right)\left(\frac{2x' + 3y'}{\sqrt{13}}\right) = 9$$

$$5(9x'^2 - 6x'y' + 4y'^2) + 12(6x'^2 + 9x'y' - 4x'y' - 6y'^2) = 117$$

$$117x'^2 - 52y'^2 = 117$$

$$x'^2 - \frac{y'^2}{\frac{9}{4}} = 1$$

23. $5x^2 + 24xy - 2y^2 = 44$

$\quad A = 5, \quad B = 24, \quad C = -2$

$\quad \cot 2\theta = \dfrac{5 - (-2)}{24} = \dfrac{7}{24}$

$\quad \cos 2\theta = \dfrac{7}{25}$

$\quad \sin \theta = \sqrt{\dfrac{1 - \frac{7}{25}}{2}} = \dfrac{3}{5}$

$\quad \cos \theta = \sqrt{\dfrac{1 + \frac{7}{25}}{2}} = \dfrac{4}{5}, \quad \theta = 36.87°$

$x = \frac{4}{5}x' - \frac{3}{5}y'$

$y = \frac{3}{5}x' + \frac{4}{5}y'$

$$5\left(\frac{4x' - 3y'}{5}\right)^2 + 24\left(\frac{4x' - 3y'}{5}\right)\left(\frac{4x' + 3y'}{5}\right) - 2\left(\frac{3x' + 4y'}{5}\right)^2 = 44$$

$$\frac{5}{25}(16x'^2 - 24x'y' + 9y'^2) + \frac{24}{25}(12x'^2 - 9x'y + 16x'y' - 12y'^2)$$
$$- \frac{2}{25}(9x'^2 + 24x'y' + 16y'^2) = 44$$

$14x'^2 - 11y'^2 = 44$

$\dfrac{x'^2}{\frac{22}{7}} - \dfrac{y'^2}{4} = 1$

25. $c^2x^2 - 8c^2x + 4y^2 + 12c^2 = 0$

$\quad c^2(x^2 - 8x) + 4y^2 = -12c^2$

$\quad c^2(x^2 - 8x + 16) + 4y^2 = -12c^2 + 16c^2$

$\quad c^2(x - 4)^2 + 4y^2 = 4c^2$

$\quad \dfrac{(x - 4)^2}{4} + \dfrac{y^2}{c^2} = 1, \quad$ ellipses

27. $y^2 + x^2 = (x + 5)^2$

$\qquad y^2 + x^2 = x^2 + 10x + 25$

$\qquad\qquad y^2 = 10x + 25, \quad$ parabola

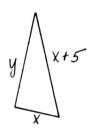

Section 10.7   Polar Equations and Their Graphs

11. $2x + 3y = 6$

$\qquad 2(r \cos \theta) + 3(r \sin \theta) = 6$

$\qquad r(2 \cos \theta + 3 \sin \theta) = 6$

13 $x^2 + y^2 - 4x = 0$

$\qquad (r \cos \theta)^2 + (r \sin \theta)^2 - 4r \cos \theta = 0$

$\qquad r^2 \cos^2 \theta + r^2 \sin^2 \theta - 4r \cos \theta = 0$

$\qquad r^2(\cos^2 \theta + \sin^2 \theta) - 4r \cos \theta = 0$

$\qquad\qquad\qquad r^2 - 4r \cos \theta = 0$

$\qquad\qquad\qquad r(r - 4 \cos \theta) = 0$

$\qquad\qquad\qquad\qquad r = 4 \cos \theta$

15. $x^2 + 4y^2 = 4$

$\qquad (r \cos \theta)^2 + 4(r \sin \theta)^2 = 4$

$\qquad r^2 \cos^2 \theta + 4r^2 \sin^2 \theta = 4$

$\qquad r^2(\cos^2 \theta + 4 \sin^2 \theta) = 4$

17. $x^2 = 4y$

$\qquad (r \cos \theta)^2 = 4r \sin \theta$

$\qquad r^2 \cos^2 \theta = 4r \sin \theta$

$\qquad r \cos^2 \theta = 4 \sin \theta$

19. $r = 5$

$\qquad \sqrt{x^2 + y^2} = 5$

$\qquad\quad x^2 + y^2 = 25$

21. $r = 10 \sin \theta$

$$\sqrt{x^2 + y^2} = \frac{10y}{\sqrt{x^2 + y^2}}$$

$$x^2 + y^2 = 10y$$

23. $r = 1 \pm 2 \sin \theta$

$$\sqrt{x^2 + y^2} = 1 + \frac{2y}{\sqrt{x^2 + y^2}}$$

$$x^2 + y^2 = \sqrt{x^2 + y^2} + 2y$$

25. $r = \dfrac{5}{1 + \cos \theta}$

$$\sqrt{x^2 + y^2} = \frac{5}{1 + \dfrac{x}{\sqrt{x^2 + y^2}}}$$

$$\sqrt{x^2 + y^2} = \frac{5\sqrt{x^2 + y^2}}{\sqrt{x^2 + y^2} + x}$$

$$1 = \frac{5}{\sqrt{x^2 + y^2} + x}$$

$$\sqrt{x^2 + y^2} + x = 5$$

$$x - 5 = \sqrt{x^2 + y^2}$$

$$(x - 5)^2 = x^2 + y^2$$

$$x^2 - 10x + 25 = x^2 + y^2$$

$$y^2 = 25 - 10x$$

31. $r = 2 \sin \theta$

θ	0	$\pi/6$	$\pi/4$	$\pi/3$	$\pi/2$	$3\pi/4$	π
r	0	1	$\frac{2}{\sqrt{2}}$	$\sqrt{3}$	2	$\frac{2}{\sqrt{2}}$	0

33. $r \sin \theta = 1$, $r = \csc \theta$

θ	0	$\pi/6$	$\pi/4$	$\pi/3$	$\pi/2$	$3\pi/4$
θ	dne	2	$\sqrt{2}$	$\frac{2}{\sqrt{3}}$	1	$\sqrt{2}$

35. $r = 1 + \sin\theta$

θ	0	$\pi/6$	$\pi/3$	$\pi/2$	$2\pi/3$	$5\pi/6$	π
r	1	$\frac{3}{2}$	$\frac{\sqrt{3}+2}{2}$	2	$\frac{\sqrt{3}+2}{2}$	$\frac{3}{2}$	1

θ	$7\pi/6$	$4\pi/3$	$3\pi/2$	$5\pi/3$	$11\pi/6$	2π
	$\frac{1}{2}$	$\frac{2-\sqrt{3}}{2}$	0	$\frac{2-\sqrt{3}}{2}$	$\frac{1}{2}$	1

37. $r = \sec\theta$

θ	0	$\pi/6$	$\pi/4$	$\pi/3$	$\pi/2$	$2\pi/3$	$5\pi/6$	π
r	1	$\frac{2}{\sqrt{3}}$	$\sqrt{2}$	2	dne	-2	$\frac{-2}{\sqrt{3}}$	-1

θ	$7\pi/6$	$4\pi/3$	$3\pi/2$	$5\pi/3$	$11\pi/6$	2π
	$-\frac{2}{\sqrt{3}}$	-2	dne	2	$\frac{2}{\sqrt{3}}$	1

39. $r = 4\sin 3\theta$

θ	$\pi/6$	$\pi/4$	$\pi/3$	$\pi/2$	$2\pi/3$	$3\pi/4$	$5\pi/6$	π
r	4	$2\sqrt{2}$	0	-4	0	$2\sqrt{2}$	4	0

41. $r = 100(1 + \cos\theta)$

$r = 1 + \cos\theta$

θ	0	$\pi/6$	$\pi/4$	$\pi/3$	$\pi/2$	$2\pi/3$	$3\pi/4$	$5\pi/6$	π
r	1	$\frac{2+\sqrt{3}}{2}$	$\frac{2+\sqrt{2}}{2}$	$3/2$	$\frac{\theta}{r}$	$1/2$	$\frac{2-\sqrt{2}}{2}$	$\frac{2-\sqrt{3}}{2}$	0

θ	$7\pi/6$	$5\pi/4$	$4\pi/3$	$3\pi/2$	$5\pi/3$	$7\pi/4$	$11\pi/6$	2π
	$\frac{2-\sqrt{3}}{2}$	$\frac{2-\sqrt{2}}{2}$	$\frac{2-\sqrt{3}}{2}$	1	$3/2$	$\frac{2+\sqrt{2}}{2}$	$\frac{2+\sqrt{3}}{2}$	2

43. $r = \dfrac{1}{2 - \cos\theta}$

$$\sqrt{x^2 + y^2} = \dfrac{1}{2 - \dfrac{x}{\sqrt{x^2 + y^2}}}$$

$$\sqrt{x^2 + y^2} = \dfrac{\sqrt{x^2 + y^2}}{2\sqrt{x^2 + y^2} - x}$$

$$2\sqrt{x^2 + y^2} - x = 1$$

$$2\sqrt{x^2 + y^2} = x + 1$$

$$4(x^2 + y^2) = (x + 1)^2$$

$$4x^2 + 4y^2 = x^2 + 2x + 1$$

$$3x^2 + 4y^2 - 2x - 1 = 0$$

Chapter 10   Review

3.
$$x^2 + y^2 + 4y = 5$$
$$x^2 + y^2 + 4y + 4 = 5 + 4$$
$$x^2 + (y + 2)^2 = 3^2$$

5.
$$(x + 1)^2 - y^2 - 2y = 0$$
$$(x + 1)^2 - (y^2 + 2y + 1) = -1$$
$$(x + 1)^2 - (y + 1)^2 = -1$$
$$(y+1)^2 - (x+1)^2 = 1$$

7.
$$x^2 - y^2 = 100$$
$$\dfrac{x^2}{100} - \dfrac{y^2}{100} = 1$$

9.
$$y^2 + 2(x^2 + 3x) = 7$$
$$y^2 + 2\left(x^2 + 3x + \dfrac{9}{4}\right) = 7 + \dfrac{9}{2}$$
$$y^2 + 2\left(x + \dfrac{3}{2}\right)^2 = \dfrac{23}{2}$$
$$\dfrac{y^2}{\dfrac{23}{2}} + \dfrac{\left(x + \dfrac{3}{2}\right)^2}{\dfrac{23}{4}} = 1$$

11. Vertex: $(2, 0)$, $a = -2$
    Focus: $(0, 0)$

$$(y - 0)^2 = 4(-2)(x - 2)$$
$$y^2 = -8(x - 2)$$

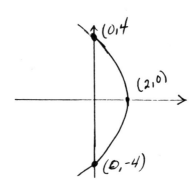

13. Center: $(-1, 2)$
    Radius: 8

$$(x + 1)^2 + (y - 2)^2 = 64$$

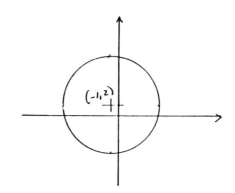

15. Center: $(0, 0)$
    Focus: $(2, 0)$, $c = 2$
    Major Axis: 8, $a = 4$

$$b^2 = 16 - 4 = 12$$
$$\frac{x^2}{16} + \frac{y^2}{12} = 1$$

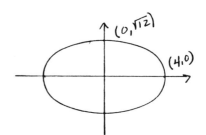

17. Vertices: $(7, 4)$, $(-1, 4)$

Asymptote: $y = \pm 2x$

Center: $(3, 4)$

$a = 4$, $b = 8$

$\dfrac{(x - 3)^2}{16} - \dfrac{(y - 4)^2}{64} = 1$

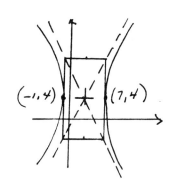

19. Major Axis: 30, $a = 15$

$b = 12$

$\dfrac{x^2}{15^2} + \dfrac{y^2}{12^2} = 1$

$c^2 = 15^2 - 12^2 = 81$, Foci: $(\pm 9, 0)$

21.

$$2x^2 + x + y^2 + 2y = 1$$

$$2\left(x^2 + \dfrac{x}{2}\right) + y^2 + 2y = 1$$

$$2\left(x^2 + \dfrac{x}{2} + \dfrac{1}{16}\right) + (y^2 + 2y + 1) = 1 + \dfrac{1}{8} + 1$$

$$2\left(x + \dfrac{1}{4}\right)^2 + (y + 1)^2 = \dfrac{17}{8}$$

$$\dfrac{\left(x + \dfrac{1}{4}\right)^2}{\dfrac{17}{16}} + \dfrac{(y + 1)^2}{\dfrac{17}{8}} = 1$$

$y + 2x = 1$, $y = -2x + 1$

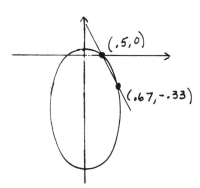

23. See p. 319 and Fig. 10.1 in the text.

25. $y = 2x$

$\frac{y}{x} = 2$

$\tan \theta = 2$

29. $4(r \cos \theta)^2 + (r \sin \theta)^2 = 1$

$4r^2 \cos^2 \theta + r^2 \sin^2 \theta = 1$

31. $r = 2$

$$\sqrt{x^2 + y^2} = 2$$

$$x^2 + y^2 = 4$$

33. $r = \dfrac{2}{1 + \sin \theta}$

$$\sqrt{x^2 + y^2} = \frac{2}{1 + \dfrac{y}{\sqrt{x^2 + y^2}}}$$

$$\sqrt{x^2 + y^2} = \frac{2\sqrt{x^2 + y^2}}{\sqrt{x^2 + y^2} + y}$$

$$\sqrt{x^2 + y^2} + y = 2$$

$$y - 2 = -\sqrt{x^2 + y^2}$$

$$(y - 2)^2 = x^2 + y^2$$

$$y^2 - 4y + 4 = x^2 + y^2$$

$$x^2 = -4(y - 1)$$

35. $r = 3 \cos \theta$

$$\sqrt{x^2 + y^2} = \frac{3y}{\sqrt{x^2 + y^2}}$$

$$x^2 + y^2 = 3y$$

$$x^2 + y^2 - 3y + \frac{9}{4} = \frac{9}{4}$$

$$x^2 + \left(y - \frac{3}{2}\right)^2 = \left(\frac{3}{2}\right)^2$$

37. $r = \sin 2\theta$

$r = 2 \sin \theta \cos \theta$

$$\sqrt{x^2 + y^2} = 2\left(\frac{y}{\sqrt{x^2 + y^2}}\right)\left(\frac{x}{\sqrt{x^2 + y^2}}\right)$$

$$(x^2 + y^2)^{3/2} = 2xy$$

39. $r^2 - r = 0$

$r(r - 1) = 0$

$r = 1$

$\sqrt{x^2 + y^2} = 1$

$x^2 + y^2 = 1$